Synthesis Lectures on Emerging Engineering Technologies

This series publishes short books on current engineering technologies that are gaining prominence, as well as promising technologies that are being developed, for an audience of researchers, advanced students, engineers and other professionals, and entrepreneurs.

Synthesis Lectures on Emerging Engineering Technologies

This series publishes short books on current engineering technologies that are gaining prominence, as well as promising technologies that are being developed, for an audience of researchers, advanced students, engineers and other professionals, and entrepreneurs.

Nabil Shovon Ashraf

Parameter-Centric Scaled FET Devices

Physics Based Perspectives and Attributes

 Springer

Nabil Shovon Ashraf
Dhaka, Bangladesh

ISSN 2381-1412   ISSN 2381-1439 (electronic)
Synthesis Lectures on Emerging Engineering Technologies
ISBN 978-3-031-84285-6   ISBN 978-3-031-84286-3 (eBook)
https://doi.org/10.1007/978-3-031-84286-3

This Springer imprint is published by the registered company Springer Nature Switzerland AG
The registered company address is: Gewerbestrasse 11, 6330 Cham, Switzerland

If disposing of this product, please recycle the paper.

Preface

This book authored by Dr. Nabil Shovon Ashraf and entitled *Parameter-Centric Scaled FET Devices* chronologically discusses in the Introduction section and Chaps. 1–8, the material and device transport parameters that are critical for FET devices at advanced node architectures such as gate all around (GAA) nanowire FET and vertically stacked GAA nanosheet FET. Most of the previous modeling of some of these parameters that define FET drive current, mobility and saturated drift velocity from T = 300 K to cryogenic temperature exclude band nonparabolicity effect on conduction band minima and valence band maxima due to high degenerate doping for silicon. This book extracts parameters such as activated dopant percentage that generates free carrier concentration, density of states effective mass for electron and hole for silicon and conductivity effective mass for electron and hole for nondegenerate doping concentrations for cryogenic temperatures up to 50 K which are done through precise analytical equations from T = 4.2 K experimental data for silicon. Since band nonparabolicity is a function of temperature distinctively for each particular doping concentration between 10^{17}/cm^3 to 10^{21}/cm^3, their derivation in analytical form requires a huge set of data from measurements or modeling at T = 300 K for silicon which has not been done so far.

Industry TCADs like Silvaco and Synopsys Quantum ATK tool report the density of states' effective mass of electron and hole values for silicon at T = 300 K with imprecise substitutions for simulation purposes. This book correctly quotes these values along with conductivity effective mass values both for nondegenerate doping of silicon substrate and non-linear incremental approach-based effect of band nonparabolicity on these two types of effective masses in silicon at T = 300 K for degenerate doping concentrations. 3D band structure-based density functional theorem (DFT) and full band 3D quantum simulation methods might generate FET transport values such as drive current, subthreshold leakage current, threshold voltage, Ion/Ioff inversion channel mobility and saturated drift velocity under ballistic transport but does not intrinsically report the density of states effective mass values or conductivity effective mass values of majority or minority carrier FETs in a precise way. The parameters generated in this book from Chaps. 1–5 can be used in TCAD simulators and other device transport analysis-based simulators like DFT, full band 3D quantum simulation and Ensemble Monte Carlo simulation techniques and benchmark parameters then will be more precisely defined as these parameters have analytical equations based on computability as narrated in Chaps. 1–5. Material properties are key for all materials where these FET devices are fabricated other than silicon material and mostly their ballistic transport efficiency is pivotal like silicon substrate in cryogenic temperatures where certain device physical analyses of these parameters become critical and are discussed in final Chap. 8. This book will enlighten and deepen the learning of device physics and engineering professionals who work on FET architectures based on device fabrication, modeling and reliability assessments at advanced nodes.

Dhaka, Bangladesh   Nabil Shovon Ashraf

Introduction

Why the author of this book notes that there are critical modeling requirements that have not been properly discussed todate by various illustrious textbooks [1–4]. Before reference [6] first showed that the general accepted theory that incomplete ionization of dopants in n-type and p-type silicon only existed as low temperatures of operation such as T = 100 K to below down to cryogenic temperature T = 4.2 K and also attributed improperly that for degenerate doping above than 10^{19}/cm^3 substrate doping in silicon for both p-type and n-type doping in silicon, incomplete ionization plays a role. Reference [6] proves these established theories wrong that generally in n-type silicon with phosphorous doping and p-type silicon with boron doping, the incomplete ionization at T = 300 K stays close to 100% between 10^{15}/cm^3 and 10^{17}/cm^3 and then the ionization gradually drops to almost 80% for n-type phosphorous doped silicon and near 70% for p-type boron doped silicon between 10^{17}/cm^3 and close to 10^{19}/cm^3. From 10^{19}/cm^3, the ionization starts to increase quickly and reaches full 100% for 10^{21}/cm^3–10^{22}/cm^3 for both n-type phosphorous and p-type boron doped silicon at T = 300 K excluding dopant values close to silicon maximum dupability limit (5×10^{22}/cm^3). So we can see that, the general accepted theory that low temperature operation generates incompletely ionized free carriers are acceptable but degenerate doping levels between few 10^{18}/cm^3 to 10^{21}/cm^3 generate incomplete ionization in silicon is wrong, rather the incomplete ionization that is experimentally demonstrated in [6] is actually in the non-degenerate doping levels (10^{17}/cm^3–few 10^{18}/cm^3) and slightly degenerate doping levels (few 10^{18}/cm^3–10^{19}/cm^3) and for general high degenerate doping level in silicon for p-type and n-type at T = 300 K, there is actually complete ionization at T = 300 K.

This observation of reported incomplete ionization modeling in [6] provides us a completely novel step by step analytical equation-based approach where before using Fermi-Dirac integral, first we need to know ionized free carrier density in silicon for n-type and p-type silicon at T = 300 K taking the Phosphorous and Boron as dopants and compute this carrier density (ionized) from provided substrate doping data or information. Correctly knowing this completely ionized free carrier density at a particular substrate doping density assuming neutral also gives us reassessment of the curve where free carrier density reduction due to complete ionization computation, changes the associated Fermi-Dirac integral value from the curve shown in [1] and the corresponding integral parameter eta_c = $(E_f\text{-}E_c)/kT$ and eta_v = $(E_v\text{–}E_f)/kT$ shown in [1]. The steps are now (1) first to determine ionized free carrier density in n-type and p-type silicon at T = 300 K taking a substrate doping [6], (2) calculate the Fermi-Dirac integral from [1] and integral parameter eta_c and eta_v from [1]. (3) Then we can verify how accurate the Fermi-Dirac integral value based on ionized free carrier computation by taking equations in [5] where knowing eta_c and eta_v corresponding Fermi-Dirac integral can be numerically calculated, and the error percentage generally remains within 0.5% for all doping levels mainly non-degenerate to degenerate transition zone ($10^{17}/cm^3$ and above) and for degenerate zone ($10^{18}/cm^3$ to $10^{21}/cm^3$).

Band nonparabolicity of density of states (DOS) effective masses for electron and hole in silicon at T = 300 K at degenerate doping levels for the values mentioned above in silicon are also not accurately reported except reference [4]. Reference [4] reports the DOS effective masses for degenerate doping levels in silicon for electron for T = 300 K and other temperatures where effective mass increase of silicon for T = 300 K for electron in the DOS case, is seen to rise above than 1.6 times than room temperature nondegenerate effective mass in silicon for electron, i.e., 1.18 m_o where m_o is free electron mass. The author also witnesses wrongly reported DOS effective mass values at T = 300 K for silicon for electron and hole. Only [1], [2] and [4] report the DOS effective mass values at T = 300 K for silicon for the case of electron and hole to be 1.18 m_o and 0.81 m_o, which is very crucial in determining many parameters for field effect transistors (FET), for instance, effective density of states in the conduction and valence band in silicon for T = 300 K which determine inversion carrier density in FET as a function of gate voltage, intrinsic carrier concentration at T = 300 K for silicon for non-degenerate doping values [11–13] taking the correctly quoted masses from [1], [2] and [4] which may not have been correctly quoted in [11–13]. Conductivity effective masses for silicon at T = 300 K for electron and hole are not also fully correctly reported by any of these textbook references [1–4]. So, the author of this book first calculated the conductivity effective masses for electron and hole in silicon for T = 300 K for non-degenerate doping levels taking analytically solvable equations from [4] and also [3]. Conductivity effective masses also need precise calculation as these determine the drift mobility of channel carriers in n-type and p-type FETs where the gate voltage and drain voltage also play role by introducing carrier scattering. So, from well-known reported mobility-doping concentration figures for majority carriers and minority inversion carriers such as [17], using Drude equation and knowing conductivity effective masses of electron and hole in silicon at T = 300 K at a specific substrate doping taking non-degenerate doping into consideration, total scattering time can be determined for low drift field. Now, band nonparabolicity also affects the conductivity effective masses of electron and hole in silicon at T = 300 K when the doping degenerates and this will modify the mobility and scattering times seen in [17] for high degenerate substrate doping where the actual mobility reduction factor is not ionized impurity scattering but surface roughness scattering but [17] only reports ionized impurity scattering-related mobility values as a function of substrate doping up to degenerate level and information about subthreshold mobility, ionized impurity scattering, thermally limited phonon scattering-related mobility peak value, transition to optical phonon scattering and surface roughness scattering can be seen from plots in [18] as a function of vertical electric field in n-MOSFET with low lateral field or drain voltage. Here also incomplete ionization modifies the ionized carriers and the screening of dopants for subthreshold mobility and surface roughness scattering, the latter should be more intense for degenerate substrate doping values in n-FET where the carriers are 100% ionized. Furthermore, [18] is inaccurate for inversion layer mobility in the doping range $10^{17}/$cm^3 to $10^{19}/$cm^3, where [18] generally assumed complete ionization for non-degenerate doping levels where values such as $10^{17}/$cm^3 to few $10^{18}/$cm^3 lie. Since, [6] shows that there is incomplete ionization of dopants in silicon at T $=$ 300 K for both n-type phosphorous and p-type boron between $10^{17}/$cm^3 to $10^{19}/$cm^3 dopant values, therefore for gate bias in subthreshold region, there will be neutral impurity scattering to be considered for inversion layer mobility in [18] which reports data for T $=$ 300 K and T $=$ 77 K. For neutral impurity scattering, for n-MOSFET, inversion electrons face less repulsive field than negatively ionized acceptors in the depletion region, as a result, neutral impurity scattering increases subthreshold mobility and inversion layer mobility for both linear and saturation region when gate overdrive is low or the phonon-related mobility peak has not been reached. The author of this book recalculated the conductivity effective masses for electron and hole in silicon for T $=$ 300 K from nondegenerate to degenerate substrate doping levels taking a ratio-based increase of masses due to band nonparabolicity. Analytical equations for DOS and conductivity effective masses for electron and hole in silicon at T $=$ 300 K up to degenerate substrate doping were very much important from FET transport parameter calculations such as inversion charge density, drift mobility, saturated drift velocity at T $=$ 300 K that determine the FET drive current and that was the sole purpose of this book that staring from accurate completely ionized dopant information, from Fermi-Dirac integral value and then corresponding eta_c and eta_v values, bulk Fermi level energy Ef can be determined, band gap narrowing from ionized dopant-related conduction band minima peak decrease and valence band maxima increase taking equations from [10] and hence the bulk potential for threshold voltage determination generally for degenerate doping substrate dopant values. Intrinsic Fermi energy level Efi for degenerate doping case can be determined from correctly computed intrinsic carrier concentration taking the references [6] and [10] and that exactly the author focused upon as the bulk potential can be known by knowing the intrinsic Fermi energy Efi and actual Fermi energy Ef.

Next the references as we go along [14–16] all need revision as per correctly computed ionized dopants from [6], changes in Fermi-Dirac integral and integral parameters eta_c and eta_v, band nonparabolicity induced density of states, effective masses for electron and hole now need to be calculated as a function of temperature and doping up to degenerate doping levels both, ionized dopant-related changes in band gap narrowing parameter as scripted [10] and making it function of temperature. Reference [19] reports the proper application rule of Matthiesen's equation to combine various scattering events in n and p-type silicon substrates and FETs. References [20–22] are very important references for scaled FET architecture overview and how 2D scaling length parameter optimizes the short channel effect (SCE) in FET but all calculations must proceed from defining ionized dopant data first and then converting it to free carriers. Fermi-Dirac integral should be used in these references [20–22], although the substrate doping in the nondegenerate regime indicates that Maxwell-Boltzmann equation use would not alter percentage error above 0.5% for all the parametric modeling data reported in these references [20–22]. Reference [23] is an important reference for scaling theory to be applied to control short channel effects (SCE) for fully depleted cylindrical and surrounding gate MOSFETs, which will be useful in studying stacked nanosheet gate all around MOS-FETs, where currently FinFET like fin type gate structure is fabricated. For thin silicon film, if fully depleted cylindrical and surrounding gate structure like gate all around GAA FET can be employed for nanosheet n-FET, using scaling theory, it can be determined how volume inversion for low gate voltage can be achieved in this nanosheet FET. Volume inversion of carriers reduces boundary layer scattering and surface scattering induced self-heating effect (SHE) that is conspicuously observed in this type of considerable fin-height and width-based gate stack.

References [24–27] focus upon gate tunneling leakage current optimization which increases at a higher rate than subthreshold leakage current as the FET nodes are scaled. Modeling and computation of these gate tunneling leakage current being also necessary for nearly 1–2 nm effective oxide thickness including high-k oxides and reduction of gate tunneling leakage current from optimizing device parameters of FET are mentioned in [24–26]. Reference [27] shows that 2D material such as MoS_2 is capable of showing much lower tunneling leakage current than silicon. The lowest interface defect density requires a seamless oxide-semiconductor interface and for that process such as atomic layer deposition (ALD) is mandatory and references [28] and [29] discuss this. References [30–35] discuss scaling effects on nanoscale MOSFET from performance benchmarking. Reference [36] is an important inclusion in this book that shows through Ensemble Monte Carlo simulation, the thermal effects of nanoscale MOSFETs, where the channel carrier temperatures can rise many kelvin over than room temperature T = 300 K setting the transport to be at non-equilibrium, and this paper also shows that heating effects get pronounced near drain before thermalization and therefore, in saturation region where the drain voltage is higher, the inversion layer carriers may undergo thermal heating effects. Boundary layer scattering and surface roughness scattering in thin silicon film such as FinFET, gate all around (GAA) nanowire FET and fully depleted silicon on insulator (FD-SOI) FET suffer additionally from self-heating effects where the temperatures of the channel carriers remain higher than substrate temperature during on-current condition as silicon's thermal conductivity is lower and cannot dissipate the heat faster than material such as SiO_2 and diamond. In FD-SOI case, self-heating effect is minimized for thin silicon buried-oxide (BOX), which has higher thermal conductivity than the silicon film that is fully depleted and generally more thinned than BOX. Thin BOX of FD-SOI also enables BOX-film interface to be having a sheet of positive potential by applying negative voltage on the back gate, so the inversion layer electrons are slightly pulled away in the silicon film by the attractive force of the BOX-silicon back gate interface sheet positive charge and inversion carriers. This reduces boundary layer scattering at the front gate oxide-silicon film interface and also surface roughness scattering for FD-SOI and hence self-heating effects (SHE).

Reference [37] is a rigorous analysis of quantum transport in field effect transistors, importantly for scaled nodes around 10 nm. References [38–42] discuss the cryogenic temperature device physics and operational characteristics of FETs. References [43–46] are based on multigate architecture all the way to gate all around (GAA) nanowire FET, which is the norm for present FET architecture due to improved gate to channel integrity, uniform channel thickness and operation region in the low nondegenerate substrate doping allowing volume inversion of channel carriers as the GAA nanowire pore dimension gate smaller. Self-heating effects (SHE) that degrade threshold voltage and FET transconductance for GAA FET with cylindrical and fin-type gates and also for reduced pore diameter are also discussed in couple of these references [43–46, 53–54]. References [47–52, 55] discuss the most recent FET architecture progressing from GAA to nanosheet with vertical gate stack and also complementary FET or C-FET form. Some of these critically important reference articles are further analyzed in various chapters of this book by the author of this book. Some factual contents regarding advanced node FETs are arbitrary and do not require particular reference citation to be indexed.

The author of this book has arranged the chapters as follows. Chapter 1 discusses all the parametric computation related to incomplete ionization in silicon at T = 300 K for the doping levels between $10^{17}/cm^3$ to few $10^{19}/cm^3$. Band nonparabolicity that changes DOS effective masses for electron and hole in silicon and associated change in incomplete ionization or ionized dopant data are also calculated and plotted with different perspectives of illustration in Chap. 2. Chapter 3 discusses the DOS and conductivity effective masses of electron and hole in silicon at T = 300 K between $10^{17}/cm^3$ and $10^{19}/cm^3$ with nondegenerate to degenerate regime and with ratioed increase of band nonparabolicity for different ranges of degenerate substrate doping. For non-degenerate substrate doping, based on full ionized dopants (doping in the range $10^{16}/cm^3$), DOS and conductivity effective masses of electron and hole as a function of substrate temperature down to 4.2 K can be determined by three-term polynomial method but these equations are not derived in Chap. 3 as band nonparabolicity effect is temperature dependent for a particular doping concentration. Analytical equation format DOS and conductivity effective masses for electron and hole in silicon derivation are complicated for higher nondegenerate to degenerate doping levels where band nonparabolicity itself becomes function of temperature distinctively for each dopant. So that calculation is omitted. Also discussed are scattering time extraction using Drude mobility model and extracted conductivity effective masses of electron and hole with band nonparabolicity for degenerate doping.

Chapter 4 discusses the Drude mobility for majority and minority carriers of electron and hole in FETs, effective intrinsic carrier concentration (n_{ieff}) for silicon at $T = 300$ K taking ionized dopant into consideration and ionized dopant induced band gap narrowing into consideration and now this n_{ieff} is shown to be different for n-type and p-type substrate and generally has a increasing trend between 10^{17}/cm^3 to few 10^{19}/cm^3. DOS masses for electron and hole for majority n-type and majority p-type silicon at $T = 300$ K are recalculated in the presence of band nonparabolicity to calculate overall n_{ieff}, other parameters like Fermi level E_F, band gap E_G, bulk potential φ_B and substrate resistivity ρ are also calculated taking the effect of band nonparabolicity on degenerate substrate doping for both n-type and p-type silicon at $T = 300$ K. Chapter 5 discusses the advantages of operating FET devices in cryogenic temperatures and some of the bottlenecks that need to be overcome for ensuring higher drive current than $T = 300$ K operation. Neutral impurity scattering which is present in silicon at $T = 300$ K for certain incompletely ionized nondegenerate doping values and also for temperatures as low as cryogenic temperatures is discussed in a sub-section of this Chap. 5. Chapter 6 discusses the current progress on GAA nanowire FET performance improvement. Chapter 7 discusses the stacked vertical nanosheet FET and one of the focus of this chapter is the self-heating effects in GAA and nanosheet FET and its minimization strategy and issues arising from random discrete channel dopants out of confinement by gate stack. Chapter 8 elucidates the conclusion on understanding of Chaps. 1–7, future directions and insights on the modeling requirement of parameters to be analytically computable for FET devices fabricated in material other than silicon at $T = 300$ K down to cryogenic temperature level operation. List of references is finally appended to this book.

References

1. Advanced Semiconductor Fundamentals, Robert F. Pierret. Volume VI, Second edition, Pearson Education Inc., 2003.
2. Semiconductor Device Fundamentals, Robert F. Pierret, Addison-Wesley Publication Company Inc, 1996.
3. Semiconductor Physics and Devices Basic Principles, Donald A. Neamen, Fourth Edition, McGraw Hill Companies Inc, 2012.
4. Low Temperature Electronics, Physics, Devices, Circuits and Applications, Edmundo A. Gutiérrez-D., M. Jamal Deen and C. Claeys, Academic Press, 2001.
5. Approximation of Fermi-dirac Integrals of Different Orders used to Determine the Thermal Properties of Metals and Semiconductors, O. N. Koroleva, A. V. Mazhukin, V. I. Mazhukin and P. V. Breslavskiy, Physics, 2016.

6. Physical Model of Incomplete Ionization for Silicon Device Simulation, Andreas Schenk, Pietro P. Altermatt and Bernhard Schmithúsen, 2006 International Conference on Simulation of Semiconductor Processes and Devices, September 2006, pp. 51–54.

7. Device Modeling at Cryogenic Temperatures: Effects of Incomplete Ionization, Akin Akturk, Jeffrey Allnutt, Zeynep Dilli, Neil Goldsman and Martin Peckerar, IEEE Transactions on Electron Devices, Volume 54, Issue 11, November 2007, pp. 2984–2990.

8. Incomplete ionization in a semiconductor and its implications to device modeling, G. Xiao, J. Lee, J. J. Liou and A. Ortiz-Conde, Microelectronics Reliability, 39, 1999, pp. 1299–1303.

9. Energy and temperature dependence of electron effective masses in silicon, Nicolas Cavassilias Jean-Luc Autran, Frédéric Aniel and Guy Fishman, Journal of Applied Physics, Volume 92, Number 3, 92, 1431 (2002), pp. 1431–1433.

10. Band gap narrowing in n-type and p-type 3C-, 2H-, 4H-, 6H-SiC, and Si, C. Persson, U. Lindefelt, and B. E. Sernelius, Journal of Applied Physics, 86, 4419 (1999), Volume 86, Number 8, pp. 4419–4427.

11. Intrinsic concentration, effective densities of states, and effective mass in silicon, Martin A. Green, Journal of Applied Physics, 67, 2944 (1990), pp. 2944–2954.

12. Reassessment of the intrinsic carrier density temperature dependence in crystalline silicon, Romain Couderc, Mohamed Amara and Mustafa Lemiti, Journal of Applied Physics, 115, 093705 (2014), pp. 093705–1 to 5.

13. Improved value for the silicon intrinsic carrier concentration from 275 to 375 K, A. B. Sproul and M. A. Green, Journal of Applied Physics, 70, 846 (1991), pp. 846–854.

14. Temperature Dependence of the Energy Gap in Semiconductors, Y. P. Varshni, Physica, 34, 1967, pp. 149–154.

15. Temperature dependence of the band gap of silicon, W. Bludau, A. Onton and W. Heinke, Journal of Applied Physics, 45, 1846 (1974), pp. 1846–1848.

16. Intrinsic carrier concentration and minority-carrier mobility of silicon from 77 to 300 K, A. B. Sproul and M. A. Green, Journal of Applied Physics, 73, 1214 (1993), pp. 1214–1225.

17. An Analytical, Temperature-dependent Model for Majority- and Minority-carrier Mobility in Silicon Devices, Susanna Reggiani, Marina Valdinoci, Luigi Colalongo, Massimo Rudan and Giorgio Baccarani, VLSI Design, January 2000, Semiconductor Device Modeling and Simulation (section), Volume 10, Number 4, pp. 467–483.

18. On the universality of inversion layer mobility in silicon MOSFET's: Part I-Effects of substrate impurity concentration, S. Takagi, A. Toriumi, M. Iwase and H. Tango, IEEE Transactions on Electron Devices, Volume 41, Issue 12, December 1994, pp. 2357–2362.

19. Error-Free Matthiessen's Rule in the MOSFET Universal Mobility Region, Ming-Jer Chen, Wei-Han Lee and Yi-Hui Huang, IEEE Transactions on Electron Devices, Volume 60, Number 2, February 2003, pp. 753–758.

20. CMOS scaling into the nanometer regime, Yuan Taur, D. A. Buchanan, Wei Chen, D. J. Frank, K. E. Ismail, Shih-Hsien Lo, G. A. Sai-Halasz, R. G. Viswanathan, H.- J. C. Wann, S. J. Wind and Hon-Sum Wong, Proceedings of the IEEE, Volume 85, Issue 4, April 1997, pp. 486–504.

21. Device scaling limits of Si MOSFETs and their application dependencies, D. J. Frank, R. H. Dennard, E. Nowak, P. M. Solomon, Y. Taur and Hon-Sum Philip Wong, Proceedings of the IEEE, Volume 89, Issue 3, March 2001, pp. 259–288.

22. Generalized Scale Length for Two Dimensional Effects in MOSFET's, David J. Frank, Yuan Taur and Hon Sum P. Wong, IEEE Electron Device Letters, Volume 19, Number 10, October 1998, pp. 385–387.

23. Scaling Theory for Cylindrical, Fully-Depleted Surrounding-Gate MOSFET's , Christopher P. Auth and James D. Plummer, IEEE Electron Device Letters, Volume 18, Number 2, February 1997, pp. 74–76.

24. Limit of Gate Oxide Thickness Scaling in MOSFETs due to Apparent Threshold Voltage Fluctuation Induced by Tunnel Leakage Current, Meishoku Koh, Wataru Mizubayashi, Kunihiko Iwamoto, Hideki Murakami, Tsuyoshi Ono, Morikazu Tsuno, Tatsuyoshi Mihara, Kentaro Shibahara, Seiichi Miyazaki and Masataka Hirose, IEEE Transactions on Electron Devices, Volume 48, Number 2, February 2001, pp. 259–264.

25. A review of gate tunneling current in MOS devices, Juan C. Ranuãrez, M. J. Deen and Chih-Hung Chen, Microelectronics Reliability, 46, 2006, pp. 1939–1956.

26. Direct-Tunneling Gate Leakage Current in Double Gate and Ultrathin Body MOSFETs, Leland Chang, Kevin J. Yang, Yee-Chia Yeo, Igor Polishchuk, Tsu-Jae King and Chenming Hu, IEEE Transactions on Electron Devices, Volume 49, Number 12, December 2002, pp. 2288–2295.

27. The Integration of Sub-10 nm Gate Oxide on MoS2 with Ultra Low Leakage and Enhanced Mobility, Wen Yang, Oing-Oing Sun, Yang Geng, Lin Chen, Peng Zhou, Shi-Jin Ding and David Wei Zhang, Scientific Reports, 5, Article Number: 11921 (2015).

28. Conformality in atomic layer deposition: Current status overview of analysis and modelling, Véronique Cremers, Riikka L. Puurunen and Jolien Dendooven, Applied Physics Reviews, 6, 021302 (2019), pp. 021302-1–021302-43.

29. Improved thermal stability and electrical properties of atomic layer deposited HfO2/AlN high-k gate dielectric stacks on GaAs, Yan-Qiang Cao, Xin Li, Lin Zhu, Zheng-Yi Cao, Di Wu and Ai-Dong Li, Journal of Vacuum Science & Technology A, 33, 01A136 (2015).

30. MOSFET Performance Scaling-Part I: Historical Trends, Ali Khakifirooz and Dimitri A. Antoniadis, IEEE Transactions on Electron Devices, Volume 55, Number 6, June 2008, pp. 1391–1400.

31. Performance Limitations of Si Bulk CMOS and Alternatives for Future ULSI, Krishna C. Saraswat, Donghyun Kim, Tejas Krishnamohan and Abhijit Pethe, Journal of The Indian Institute of Science, Volume 87:3, July–September 2007, pp. 387–400.

32. On the scaling of subnanometer EOT gate dielectrics for ultimate nano CMOS technology, Hei Wong and Hiroshi Iwai, Microelectronic Engineering, 138, (2015), pp. 57–76.

33. Technology and Modeling of Nonclassical Transistor Devices, George V. Angelov, Dimitar N. Nikolov and Marin H. Hristov, Journal of Electrical and Computer Engineering, Volume 2019, Article ID 4792461, pp. 1–18.

34. In Quest of the "Next Switch": Prospects for Greatly Reduced Power Dissipation in a Successor to the Silicon Field-Effect Transistor, Thomas N. Theis and Paul M. Solomon, Proceedings of the IEEE, Volume 98, Number 10, December 2010, pp. 2005–2014.

35. Can 3-D Devices Extend Moore's Law Beyond 32 nm Technology Node? Marius Orlowski and Andreas Wild, ECS Transactions, 3, (6), (2006), pp. 3–17.

36. Heating Effects in Nanoscale Devices, Dragica Vasileska, Katerina Raleva and Stephen M. Goodnick, Cutting Edge Nanotechnology (InTech Publisher book), Dragica Vasileska (Editor), Chapter 3, pp. 33–59.

37. A review of quantum transport in field-effect transistors, David K. Ferry, Josef Weinbub, Mihail Nedjalkov and Siegfried Selberherr, Semiconductor Science and Technology, 37 (2022), pp. 1–32.

38. An Empirically Validated Virtual Source FET Model for Deeply Scaled Cool CMOS, Wriddhi Chakraborty, Kai Ni, Jeffrey Smith, Arijit Raychowdhury and Suman Datta, IEDM 19-958, pp. 39.4.1–39.4.4.

39. Physics and performance of nanoscale semiconductor devices at cryogenic temperatures, F. Balestra and G. Ghibaudo, Semiconductor Science and Technology, 32, (2017), 023002, pp 1–14.

40. Analytical Modeling of Cryogenic Subthreshold Currents in 22-nm FDSOI Technology, Hung-Chi Han, Zhixing Zhao, Steffen Lehmann, Edoardo Charbon and Christian Enz IEEE Electron Device Letters, Volume 45, Issue 1, January 2024, pp. 92–95.

41. Theoretical Limit of Low Temperature Subthreshold Swing in Field-Effect Transistors, Arnout Beckers, Farzan Jazaeri and Christian Enz, IEEE Electron Device Letters, Volume 41, Issue 2, February 2020, pp. 276–279.
42. Cryogenic Subthreshold Swing Saturation in FD-SOI MOSFETs described with Band Broadening, H. Bohuslavskyi, A. G. M. Jansen, S. Barraud, V. Barral, M. Cassé, L. Le Guevel, X. Jehl, L. Hutin, B. Bertrand, G. Billiot, G. Pillonet, F. Arnaud, P. Galy, S. De Franceschi, M. Vinet and M. Sanquer, IEEE Electron Device Letters, Volume 40, Issue 5, May 2019, pp. 784–787.
43. Multigate transistors as the future of classical metal-oxide-semiconductor field-effect transistors, Isabelle Ferain, Cynthia A. Collinge and Jean-Pierre Collinge, Nature, Volume 479, 17 November 2011, pp. 310–316.
44. An Insight into Self-Heating Effects and Its Implications on Hot Carrier Degradation for Silicon-Nanotube-Based Double Gate-All-Around (DGAA) MOSFETs, Arun Kumar, P. S. T. N. Srinivas and Pramod Kumar Tiwari, IEEE Journal of the Electron Devices Society, Volume 7, 16 October 2019, pp. 1100–1108.
45. Physical Insight into Self-Heating Effects in Ultrathin Junctionless Gate-All-Around FETs, Arun Kumar, P. S. T. N. Srinivas and Pramod Kumar Tiwari, 2019 IEEE 9th International Nanoelectronics Conferences (INEC), 3–5 July 2019, pp. 1–4.
46. Recent Developments in Negative Capacitance Gate-All-Around Field Effect Transistors: A Review, Laixiang Qin, Chunlai Li, Yiqun Wei, Guoqing Hu, Jingbiao Chen, Yi Li, Caixia Du, Zhangwei Xu, Xiumei Wang and Jin He, IEEE Access, Volume 11, 9 February 2023, pp. 14028–14042.
47. NS-GAA FET Compact Modeling: Technological Challenges in Sub-3-nm Circuit Performance, Fabrizio Mo, Chiara Elfi Spano, Yuri Ardesi, Massimo Ruo Roch, Gianluca Piccinini and Marco Vacca, Electronics 2023, 12, 1487, pp. 1–14.
48. Compact Model for Geometry Dependent Mobility in Nanosheet FETs, Avirup Dasgupta, Shivendra Singh Parihar, Harshit Agarwal, Pragya Kushwaha, Yogesh Singh Chauhan and Chenming Hu, IEEE Electron Device Letters, Volume 41, Issue 3, March 2020, pp. 313–316.
49. A Review of the Gate-All-Around Nanosheet FET Process Opportunities, Sagarika Mukesh and Jingyun Zhang, Electronics, 2022, 11, 3589, pp. 1–11.
50. Design Insights of Nanosheet FET and CMOS Circuit Applications at 5-nm Technology Node, V. Bharath Srinivasulu and Vadthiya Narendar, IEEE Transactions on Electron Devices, Volume 69, Number 8, August 2022, pp. 4115–4122.
51. A Review of Reliability in Gate-All-Around Nanosheet Devices, Miaomiao Wang, Micromachines, 2024, 15, 269, pp. 1–20.
52. Core-insulator embedded nanosheet field-effect transistor for suppressing device-to-device variations, Donghwi Son, Hyunwoo Lee, Hyunsoo Kim, Jae-Hyuk Ahn and Sungho Kim, Scientific Reports, 14, Article Number: 7462 (2024), pp. 1–7.
53. Performance Limit of Gate-All-Around Si Nanowire Field-Effect Transistors: An Ab Initio Quantum Transport Simulation, Shiqi Liu, Quihui Li, Chen Yang, Jie Yang, Lin Xu, Linqiang Xu, Jiachen Ma, Ying Li, Shibo Fang, Baochun Wu, Jichao Dong, Jinbo Yang and Jing Lu, Physical Review Applied, 18, 054089, 2022, pp. 054089-1–054089-15.
54. Size-Dependent-Transport Study of In0.53Ga0.47As Gate-All-Around Nanowire MOSFETs: Impact of Quantum Confinement and Volume Inversion, Jiangjiang J. Gu, Heng Wu, Yiqun Liu, Adam T. Neal, Roy G. Gordon and Peide D. Ye, IEEE Electron Device Letters, Volume 33, Issue 7, July 2012, pp. 967–969.
55. Demonstration of a Nanosheet FET With High Thermal Conductivity Material as Buried Oxide: Mitigation of Self-Heating Effect, Sunil Rathore, Rajeewa Kumar Jaisawal, P. N. Kondekar and Navjeet Bagga, IEEE Transactions on Electron Devices, Volume 70, Issue 4, April 2023, pp. 1970–1976.

Contents

Dr. Nabil Shovon Ashraf was born in Dhaka, Bangladesh on August 10, 1974. Dr. Ashraf served as an Associate Professor in the Department of Electrical and Computer Engineering of North South University, Dhaka, Bangladesh from April 2018 till May 2022. From September 2014 till April 2018, he served as a faculty in the Department of Electrical and Computer Engineering of North South University Dhaka, Bangladesh with the rank of Assistant Professor. He obtained a Bachelor of Technology degree in Electrical Engineering from Indian Institute of Technology Kanpur, India in 1997. He obtained a Master of Science degree in Electrical Engineering from University of Central Florida Orlando, USA in 1999. He obtained a Doctor of Philosophy degree in Electrical Engineering from Arizona State University Tempe, USA in 2011. From December 2011 to May 2014, he was a Postdoctoral Researcher in the Department of Electrical Engineering of Arizona State University Tempe. He was employed as design engineer in RF Monolithics Inc., a surface acoustic wave (SAW)-based filter design company in Dallas, Texas, USA from August 1999 to March 2001. From October 2003 till June 2006, he served on the faculty as Assistant Professor of the Department of Electrical and Electronic Engineering of Islamic University of Technology, Gazipur, Bangladesh. Up to December 2024, Dr. Ashraf has published eight peer-reviewed journal articles (two IEEE EDS) including another in Silicon, a Springer Nature journal article in January 2024, a Journal of Nanotechnology and Nanomaterials (Scientific Archives)

journal article and about 15 international conference proceedings (three IEEE EDS). He is the author of two books on semiconductor device scaling by Morgan & Claypool publishers (Released in 2018 and 2016) and now under Springer Nature publication since January 2022, on substrate temperature engineering-based device performance improvement. Dr. Ashraf has contributed to eight book chapters on two being on interface trap-induced threshold voltage fluctuations in the presence of random channel dopants of scaled n-MOSFET at the invitations of highly accomplished international book editors and six solely-authored Chapters in *Handbook of Emerging Materials For Semiconductor Industry*, a Springer Nature Handbook published in June 2024. Together including the separate book chapters from his published books, a total of 21 book chapters have been published. He was cited in Who's Who in America for Marquis online biographies of distinguished and eminent researchers for 2 consecutive years 2015 (69th edition) and 2016 (70th platinum edition). In 2017, Dr. Ashraf became the recipient of Albert Nelson Marquis Lifetime Achievement Award honored by Marquis Who's Who. Dr. Nabil Shovon Ashraf served as a reviewer for a number of MDPI-published journals and reviewed one article from Journal of Applied Physics journal. He specializes in the area of device physics and modeling analysis of scaled devices for enabling improved device performance at the scaled node of current MOSFET device architectures, advanced lithographic techniques including high-NA EUV lithography, nanoimprint lithography and block co-polymer (BCP)-based directed self-assembly (DSA)-based advanced FET fabrication with 2–3 nm nodes.

Incomplete Ionization Related Parameter Calculations and Device Physical Effects Without Considering Band Non Parabolicity Effect at T = 300 K

Preview of the this chapter:

Incomplete ionization percentage calculation and actual ionized free carrier concentration in n-type silicon for Phosphorous doping and p-type silicon for Boron doping for T = 300 K assuming (1) non degenerate to degenerate doping values without band non parabolicity effect and (2) non degenerate to degenerate doping values with band non parabolicity effect of density of states (DOS) effective mass increase of electron and hole in silicon as a function of doping which actually starts from non degenerate doping values near slightly above than 10^{17}/cm^3. (Case (2) is discussed in Chap. 2).

The way to proceed to calculate ionized impurity carriers, is not from generally accepted equation form with Fermi-Dirac integral being calculated first and substituted in the equation $n_o = N_c \mathfrak{F}(\eta_c)$ where n_o is free carrier density (ionized), N_c is effective density of states where DOS electron effective mass at T = 300 K is assumed constant at 1.18 m$_o$ [1] and $\mathfrak{F}(\eta_c)$ is Fermi-Dirac Integral value at $\eta_c = \frac{E_F - E_c}{kT}$ where E$_F$ is extrinsic Fermi energy level (eV) and E$_C$ is the minima of conduction band (eV) and E$_C$ value gradually decreases from its fixed value 1.12 eV as a result of band gap narrowing which sets at a value slightly above than 10^{17}/cm^3 to degenerate doping levels up to 10^{21}/cm^3 where again band gap narrowing has to be calculated from provided equations such as in [2] which is most accurate to-date as per the author of this book's viewpoints assuming E$_V$ (maxima of valence band) as a reference 0 eV). So, we need to choose a different step by step extraction of parameters such as ionized free carriers, then compute $\mathfrak{F}(\eta_c)$ and then $\eta_c = \frac{E_F - E_c}{kT}$ as now ionized N$_D^+$ that goes into determining band gap narrowing and dip of E$_C$ is known from first ionization calculation sequence. From [3] for n-type silicon with Phosphorous doping at T = 300 K, the following analytical equations should be used in sequence to calculate N$_D^+$ and N$_D^+$/N$_D$ ratio and percentage and this N$_D^+$ is

N. S. Ashraf, *Parameter-Centric Scaled FET Devices*, Synthesis Lectures on Emerging Engineering Technologies, https://doi.org/10.1007/978-3-031-84286-3_1

essentially the free carrier concentration n_o that has been shown here. The equations are :

$$b(N_D) = \left\{ 1 + \left(\frac{N_D}{6 \times 10^{18}} \right)^{2.3} \right\}^{-1} \tag{1.1}$$

$$E_{dop} = \frac{45.5 \times 10^{-3}}{1 + \left(\frac{N_D}{2.2 \times 10^{18}} \right)^2} \tag{1.2}$$

$$n_1 = N_c e^{-\frac{E_{dop}}{kT}} \tag{1.3}$$

$$\frac{N_D^+}{N_D} = 1 - \left\{ b(N_D) \left(\frac{N_D^+}{(N_D^+ + gn_1)} \right) \right\} \tag{1.4}$$

Equation (1.4) in [3] is an implicit equation whose explicit form to extract N_D^+ is not given in [3]. So the author provides the quadratic Eq. (1.5) below reworked from equation (1.4) above whose solution must be used with the root that gives $N_D^+ > 0$ and also $\leq N_D$. The equation (1.5) is given below:

$$\left(N_D^+ \right)^2 + N_D^+ (gn_1 - N_D + (b(N_D)N_D)) - gn_1 N_D \tag{1.5}$$

Using [3] for actual substrate donor doping for n-type silicon with phosphorous N_D, some parametric values shown in Eqs. (1.1) to (1.5) are constant empirically fitted values from device physical perspectives for phosphorous as n-type dopant in silicon as per [3] for T = 300 K and also for other temperatures down to T = 4.2 K and we generally know that higher temperatures than T = 300 K, doping values up to degenerate levels in n-type and p-type silicon for all dopants are fully ionized and high temperature device physical modeling based parameter extraction is not the subject of this book.

The author now provides the actual ionized carriers and ionization percentage in silicon for phosphorous dopant at T = 300 K excluding degenerate doping induced band non parabolicity effect. The doping values are selected in such way so that these values represent the non degenerate doping values at the higher end to degenerate doping values near 10^{19}/cm^3.

Table 1.1 shows that the ionization percentage dopants is not 100 % for all non-degenerate doping level values and is decreasing towards the higher end of non degenerate values but the ionization percentage quickly increases to near 100% as doping becomes more degenerate and with band non parabolicity as will be shown later in another Chapter, that this ionization reaches to more higher percentage values when band non parabolicity effect is considered for both n-type silicon with phosphorous doping and p-type silicon with boron doping. The following Fig. 1.1 at T = 300 K shows the ionization percentage plotted as a function of substrate doping for n-type silicon for phosphorous doping at T = 300 K without band non parabolicity effect:

Table 1.1 Actual ionization values and percentage of ionization in n-type silicon at T = 300 K for phosphorous doping without band non parabolicity effect

N_D (/cm^3)	N_D^+ (/cm^3)	N_D^+/N_D (%)
2.18×10^{17}	2.033×10^{17}	93.257
1.61×10^{18}	1.3×10^{18}	80.75
5.34×10^{18}	4.535×10^{18}	84.925
1.1×10^{19}	1.012×10^{19}	92.00
1.69×10^{19}	1.6173×10^{19}	95.698
2.19×10^{19}	2.129×10^{19}	97.215
2.56×10^{19}	2.5063×10^{19}	97.90
2.81×10^{19}	2.7605×10^{19}	98.238
2.97×10^{19}	2.923×10^{19}	98.418
3.07×10^{19}	3.024×10^{19}	98.50

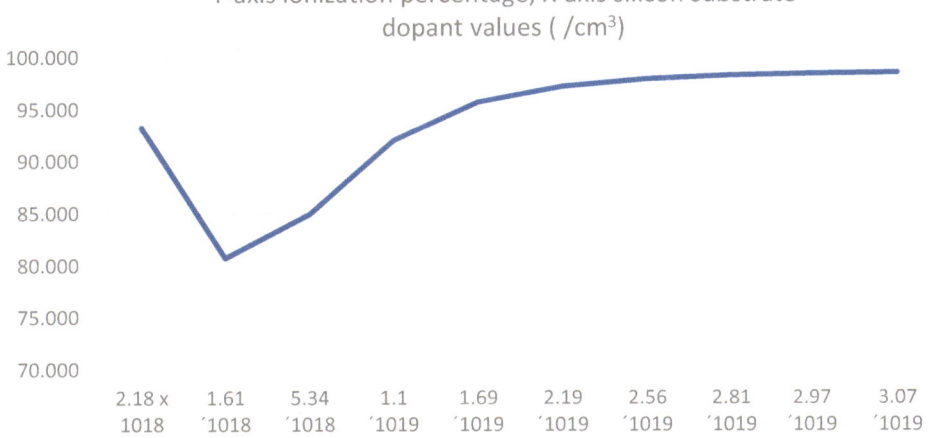

Fig. 1.1 Excel trend plot for actual ionized dopant percentage for phosphorous in n-type silicon at T = 300 K without considering band non parabolicity effect. 1017, 1018 and 1019 data sequences as seen in the X-axis doping labels are actually numbered dopant values multiplied with 10^{17}, 10^{18}, 10^{19} as per the format we see in Table 1.1 dopant values

Device physics based explanation of the shape of curve trend for Fig. 1.1 showing that dopants are generally defined by textbook incomplete ionization [4] based ionized dopants near 10^{17}/cm^3 but after some value near 1.6×10^{18} /cm^3 when the depleted ionized carriers start to bend up and up to few 10^{18}/cm^3 to 10^{19}/cm^3, the rate of ionization increase is faster and above 10^{19}/cm^3, the rate gradually increases to 100 % which is defined as degenerate doping region 10^{19}/cm^3 to 10^{21}/cm^3. Ideally, the donor activation energy value E_D (eV) through which [4] calculates the non-degenerate to degenerate doping regime incomplete ionization, E_D levels are discrete and is usually 0.045 eV assumed for both Boron and Phosphorous in silicon at T = 300 K. This is a wrong theory when degenerate regime many-body interactions are encountered that necessitates that all calculations of device parameters have to be done based on device at non-equilibrium. Furthermore, the discrete spacing of E_D levels is so minute when N_D is near 10^{19}/cm^3 or above, that E_D levels are indistinguishable from one another and forms a band that can over lap with minima of conduction band energy level E_C which also bends down due to ionized doping dependent band gap narrowing taking the case of n-type silicon at T = 300 K. As a result, for high degenerate doping, the incomplete ionization computation equation [4] cannot be used and alternative precise analytical equations like in [3] have to be used to incorporate this device physical effects in the band structure of n-type silicon for phosphorous doping at T = 300 K. [5] uses incomplete ionization calculation equation in [4] in a different way. First [5] models E_D as a function of substrate donor doping where the equation provided in [5] shows that E_D gradually moves closer to E_C as doping becomes increased to degenerate doping level, but solving in this way can produce error in final equation format provided in [5] from several other empirical parameters as the indistinguishability of E_D, the device physical attribute of many-body exchange relationship out of non-equilibrium in silicon device, should not allow any single E_D value to be used in similar like equation format given in [4]. In [3], E_D-donor dependence issue is bypassed by Eq. (1.2) and actual ionization is calculated from a step by step minimally configured empirical parameters based quadratic equation [6], which does not suffer from complicated requirements from non equilibrium many body exchange relationship based scenario. In [7] from Fig. 1.1, although the authors used Fermi-Dirac integral, they constructed equations in the format presented in [4] and hence, computed values show at T = 300 K, when $N_D = 10^{19}$/cm^3, ionization is near 40% and at $N_D = 10^{20}$/cm^3, ionization is near less than 20%, so all these data are wrong beyond the non-degenerate doping levels which are generally solved by equations in [4] and this shows the accuracy and precision of [3] that has been discussed in this chapter and the Eqs. (1.1)–(1.5) are used to rightfully demonstrate the carrier ionization values as shown in Table 1.1 and Fig. 1.1 in this chapter taking device physics into consideration for ionization assessment. The values provided in Table 1.1 and for other substrate doping as long as ionization is computed by combining (1.1)–(1.5) in this chapter, can be used for phosphorous doped n-type silicon substrate at T = 300 K for p-FET device performance analysis where holes form the channel carriers at inversion. P-FETs are not researched well for silicon and their effective mobilities are also lower than n-FET in silicon at T = 300 K. Therefore, the Table 1.1 data can be used to compute the threshold voltage of p-FET by properly calculating bulk potential which itself at T = 300 K is function of ionized donor doping density and effective intrinsic carrier concentration taking the effect of band gap narrowing which will be shown in a subsequent another Chapter. For junctionless FET type device, junctionless n-FET which is majority n-type carrier device FET in silicon at T = 300 K, Table 1.1 data are also useful as the carriers in transport from source to drain comes directly from ionized donors which is what Table 1.1 and Fig. 1.1 shows here. These ionized donors also are important from drift based mobility induced drain current at high lateral field, as majority carrier ionized impurity scattering are important here and for high ionized doping concentration near 100% for 10^{20}/cm^3 to 10^{21}/cm^3 substrate doping for n-type silicon at T = 300 K, the majority n-type ionized carriers channel is thicker than normal n-FET built on p-substrate with higher drain contact bias, the result is that interface roughness scattering becomes less in junctionless n-FET at T = 300 K for such high majority ionized carrier density with more uniform higher channel thickness, although as will be discussed in later chapter related to scattering time extractions for majority carrier mobility and minority carrier mobility of inversion channels in n-FET and p-FET, in junctionless FET as degenerate doping of the substrate close to 10^{21}/cm^3 is a must for both n-type and p-type silicon at T = 300 K, for n-type junctionless FET, the ionized carriers are electrons and donors are positively charged, so electrons see an attractive force-field towards ionized donors resulting in more direct collision and momentum loss leading to more ionized impurity scattering based mobility reduction than what happens in a minority electron and ionized acceptor based ionized impurity scattering, where scattering is not so intense as there is a repulsive force-field between inversion layer electron and depletion region ionized acceptors both being negatively charged in n-FET. But at high drain bias and lateral field, junctionless n-FET gains in mobility more than n-FET in conventional structure which is explained due to more majority carrier electron channel concentration and thickness in junctionless n-FET compared to conventional n-FET, which results in reduced interface roughness scattering for junctionless n-FET over conventional n-FET.

For p-type silicon at T = 300 K, the band structure for hole is complicated by the fact, that it has light hole mass band, heavy hole mass band and a split-off band and therefore, even though boron dopants in p-type silicon at T = 300 K, has an activation acceptor energy E_A = 0.045 eV as activation donor energy level E_D phosphorous dopant in n-type silicon at T = 300 K, the incomplete ionization and hence completely ionized dopants for the case of boron doping in p-type silicon at T = 300 K, has reduced values than Table 1.1 quoted values for ionized dopants, only at degenerate doping levels, both boron doping in p-type silicon and phosphorous doping in n-type silicon at T = 300 K, show near identical values owing to more effective mass increase of hole at T = 300 K in p-type silicon due to non parabolicity at the high degenerate doping range 10^{20}/cm^3 to 10^{21}/cm^3. The Eqs. (1.1)–(1.5) empirical constants as given in [3] for phosphorous dopant in n type silicon are now different as per [3] for boron doping in p-type silicon, The author therefore, lists these equation sequences for computation of ionized dopants in p-type silicon for boron at T = 300 K following the same process and method as deliberated for n-type silicon phosphorous dopant case.

$$b(N_A) = \left\{ 1 + \left(\frac{N_A}{4.5 \times 10^{18}} \right)^{2.4} \right\}^{-1} \qquad (1.6)$$

Table 1.2 Actual ionization values and percentage of ionization in p-type silicon at T = 300 K for boron doping without band non parabolicity effect

N_A (/cm^3)	$N_A{}^-$ (/cm^3)	$N_A{}^-/N_A$ (%)
2.18×10^{17}	1.845×10^{17}	84.63
1.61×10^{18}	1.1125×10^{18}	69.099
5.34×10^{18}	4.2089×10^{18}	78.82
1.1×10^{19}	1.0185×10^{19}	92.59
1.69×10^{19}	1.6366×10^{19}	96.84
2.19×10^{19}	2.1502×10^{19}	98.18
2.56×10^{19}	2.5270×10^{19}	98.71
2.81×10^{19}	2.7805×10^{19}	98.95
2.97×10^{19}	2.9425×10^{19}	99.074
3.07×10^{19}	3.0436×10^{19}	99.14

$$E_{dop} = \frac{44.39 \times 10^{-3}}{1 + \left(\frac{N_D}{1.3 \times 10^{18}}\right)^{1.4}} \tag{1.7}$$

$$p_1 = N_V e^{-\frac{E_{dop}}{kT}} \tag{1.8}$$

$$\frac{N_A^-}{N_A} = 1 - \left\{ b(N_A)\left(\frac{N_A^-}{\left(N_A^- + gp_1\right)}\right) \right\} \tag{1.9}$$

$$\left(N_A^-\right)^2 + N_A^-(gp_1 - N_A + (b(N_A)N_A)) - gp_1N_A \tag{1.10}$$

For N_V, effective density of states in the valence band for silicon at T = 300 K, DOS hole effective mass m_p is 0.81 m_0 [1] which is fixed without considering degeneracy induced band non parabolicity effect. Now for the same substrate data set values, Table 1.2 shows the substrate acceptor dopant concentration for boron in p-type silicon, ionized acceptor dopant concentration and ionization percentage.

Figure 1.2 shows the ionization percentage of boron dopant as acceptors in p-type silicon at T = 300 K. Please note that same silicon substrate doping concentration values as n-type dopant for the case of phosphorous as in Table 1.1, is used here in Table 1.2 and for the plot of Fig. 1.2.

As is evident from Fig. 1.2 that for boron doped p-type silicon at T = 300 K, the incomplete ionization is more intense leading much lower ionized dopants for 1.61×10^{18}/cm^3 substrate dopant but for p-type silicon with boron doping at T = 300 K, the rise to 100% ionization is much steeper as Fig. 1.2 shows. This is the reason that for degenerate doping near 10^{19}/cm^3 and higher up to 10^{21}/cm^3, p-type silicon for boron doping results in more ionized dopants at T = 300 K than its counterpart n-type silicon with

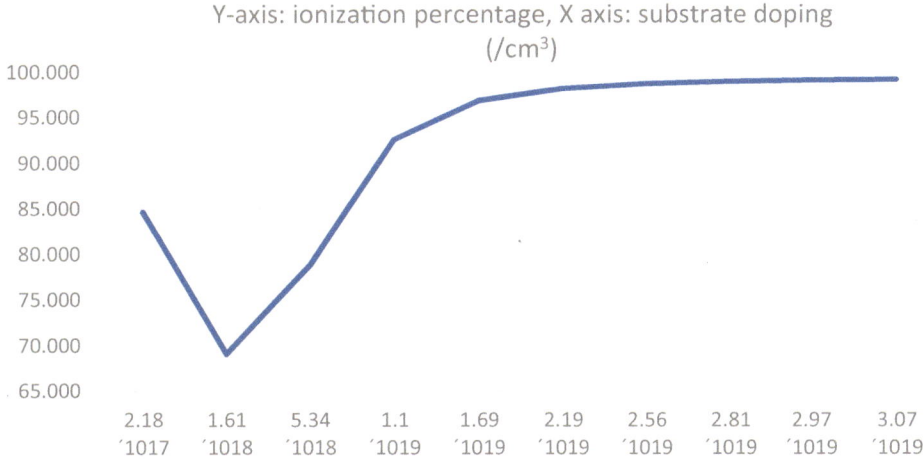

Fig. 1.2 Excel trend plot for actual ionized dopant percentage for boron in p-type silicon at T = 300 K without considering band non parabolicity effect. 1017, 1018 and 1019 data sequences as seen in the X-axis doping labels are actually numbered dopant values multiplied with 10^{17}, 10^{18}, 10^{19} as per the format we see in Table 1.1 dopant values

phosphorous dopant in the same doping range. Therefore n-FET built with such degenerately doped with boron silicon substrate, threshold voltage will be higher than p-FET where n-type silicon substrate with phosphorous dopant is used in the same degenerate doping range. Benefits of incomplete ionization is that, depletion region becomes wider and thermally generated electrons in n-FET and holes for p-FET add to the inversion carriers electron in n-FET and holes in p-FET where for both FETs, source to substrate barrier at the source end is reduced due to reduced built in voltage owing to partially ionized dopants and source to channel electron injection for n-FET and hole injection for p-FET becomes higher for same gate to source voltage. Due to much reduced ionization in p-type silicon for boron at T = 300 K, n-FET built in this substrate with non degenerate doping up to 10^{17}/cm^3, will have lower V_{bi} or built-in voltage for the same doping in n-type silicon by phosphorous at T = 300 K. So, n-FET therefore will benefit for this lower ionized dopants as source to channel electron injection will be higher for same gate to source voltage. Also, as threshold voltage is reduced, for both FET, more inversion carriers mean more screening of ionized impurity dopants and reduced ionized impurity scattering and slightly lower interface roughness scattering where for both case, n-FET will benefit more as explained above. But incomplete ionization near 1.61×10^{18} /cm^3 can result in significant neutral impurity concentrations (more so for p-silicon substrate) and this may cause randomness of ionized dopants in the depletion region of both n-FET and p-FET, channel may be slightly inhomogeneous with apparent discontinuity, inhomogeneous ionized impurity screening and inhomogeneous surface roughness scattering. Also, additional neutral impurity scattering due to incomplete ionization in the doping

Table 1.3 Actual ionization values and ionization ratio in p-type silicon with respect to n-type silicon at T = 300 K for boron doping (p-type silicon) and phosphorous doping (n-type silicon) without band non parabolicity effect

N_A^- (/cm^3)	N_D^+ (/cm^3)	N_A^-/N_D^+
1.845×10^{17}	2.033×10^{17}	0.9075
1.1125×10^{18}	1.3×10^{18}	0.8558
4.2089×10^{18}	4.535×10^{18}	0.9280
1.0185×10^{19}	1.012×10^{19}	1.0064
1.6366×10^{19}	1.6173×10^{19}	1.012
2.1502×10^{19}	2.129×10^{19}	1.00996
2.5270×10^{19}	2.5063×10^{19}	1.0083
2.7805×10^{19}	2.7605×10^{19}	1.0072
2.9425×10^{19}	2.923×10^{19}	1.0067
3.0436×10^{19}	3.024×10^{19}	1.0065

ranges shown in Tables 1.1 and 1.2, shift the total scattering time to be lower due to Matthiesen's rule and this will affect subthreshold mobility and inversion channel mobility of both n-FET and p-FET. These anomalous effects gradually vanish as dopants in the substrate is increased to degenerate doping values for which both n-type and p-type silicon at T = 300 K show near complete ionization trend and randomness of dopants in the depletion region underneath the channel gradually dies down as there is more ionization and less to lesser neutral impurity concentrations. Table 1.3 and Fig. 1.3 show the N_A^-/N_D^+ and N_A^-/N_D^+ for the same substrate doping range and values of N_A for p-type silicon and N_D for n-type silicon at T = 300 K for the case of boron and phosphorous doping.

Many a times semiconductor manufactures want to compute what should be the accurate substrate dopant concentration in n-type silicon or p-type silicon at T = 300 K. This information is important from device physical aspects for interpretation of FET performance such as (i) lower substrate resistivity to prevent latch-up due to parasitic bipolar transistor turn on effect from the elevated base potential voltage where the base is actually the substrate and this happens in both in n-FET and p-FET where drain current suddenly rises to high value with drain to source voltage latched on to a very low voltage that can burn the FET, (ii) substrate doping is also important for minimum substrate leakage current that flows out of the depletion region when the FET is biased to operate and higher substrate doping density reduces the depletion region width and subsequently increases the reverse electric field that gives rise to substrate current. This substrate current along with substrate resistance between the edge of maximum extent of depletion region and substrate contact, also modulates the actual potential that is raised from the typically grounded substrate contact considering if moderate substrate current flows through the substrate resistance that exists between the edge of maximum extent of depletion region and substrate contact. If relative positioning of the Fermi energy level E_F is known with respect to E_C (conduction band minima) in n-type silicon, then knowing $\eta_c = \frac{E_F - E_c}{kT}$, first

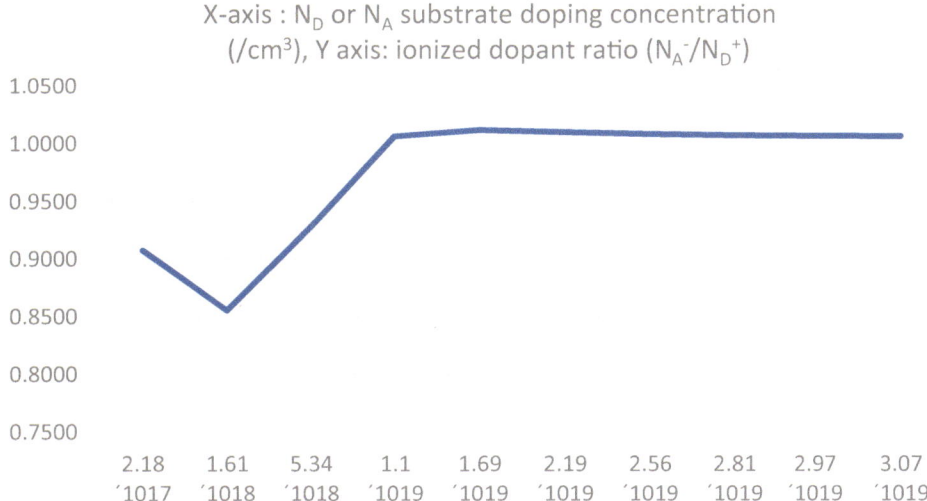

X-axis : N_D or N_A substrate doping concentration
(/cm³), Y axis: ionized dopant ratio (N_A^-/N_D^+)

Fig. 1.3 The ionization ratio N_A^-/N_D^+ as a function of actual substrate doping either in p-type silicon with boron doping or n-type silicon with acceptor doping at T = 300 K. The point to notice is after initial decrease (less ionized dopants for p-silicon), the curve rises quickly for still non degenerate doping regime showing variation in ionization in the two substrates and then for large enough non degenerate to degenerate doping, the ratio stays close to 1 and slightly above than 1, showing p-type ionization in this dopant regime is minutely higher than n-type ionization

one needs to calculate the Fermi-Dirac integral $\mathfrak{F}(\eta_c)$ from a series of equations given in [6] that is mathematically accurate to within 0.5% from actual numerically calculated value from known programs like C and C++. These equations are listed below as per [6]:

$$v(\eta_c) = \eta_c^4 + 50 + 33.6\eta_c\left\{1 - 0.68\exp\left[-0.17(\eta_c + 1)^2\right]\right\} \tag{1.11}$$

$$\varphi(\eta_c) = \frac{3\sqrt{\pi}}{4v(\eta_c)^{3/8}} \tag{1.12}$$

$$\mathfrak{F}(\eta_c) = \left[e^{-\eta_c} + \varphi(\eta_c)\right]^{-1} \tag{1.13}$$

Then using m_n, DOS effective mass of electron of silicon at T = 300 K as 1.18 m_0 [1] where m_0 is free carrier mass, the ionized dopant or free electron carrier concentration can be determined from:

$$n_o = N_c\mathfrak{F}(\eta_c) \tag{1.14}$$

This n_o is ideally N_D^+ that has been calculated through previous Eqs. (1.1)–(1.5). Putting this N_D^+ in Eq. (1.5) with some other component of that equation to be explicitly calculable and some other components being function of N_D, the substrate doping density, we appear at an implicit form of equation

$$f(N_D) = c \qquad (1.15)$$

In (1.16), f (N_D) is a multiple number terms equation involving functions of N_D whereas c is a calculated term from (1.5) taking help of (1.1)–(1.4) involving empirical parameters for phosphorous doping in n-type silicon using [3]. Now by trial and error method, (1.16) can be solved to compute N_D. A beforehand knowledge of actual ionization percentage from figure like Fig. 1.1 and tabulated values like Table 1.1, initial guess of N_D can reduce the number of iterations to arrive at correct N_D or donor substrate doping concentration in n-type silicon at T = 300 K. For p-type silicon with boron doping, Eqs. (1.11)–(1.13) are same and now $\eta_v = \frac{E_v - E_F}{kT}$ and (1.14) is replaced by analogous,

$$p_o = N_v \mathfrak{F}(\eta_v) \qquad (1.16)$$

where, p_o for p-type silicon is N_A^- as ionized acceptor dopant concentration at T = 300 K for silicon. Putting N_A^- in Eq. (1.10) and the different empirical constants related functions of N_A Eqs. (1.6)–(1.9), we arrive at a similar equation form like (1.15), i.e.,

$$f(N_A) = d \qquad (1.17)$$

Following the same method as enunciated to solve (1.15), Eq. (1.17) can be solved to determine acceptor substrate concentration in p-type boron doped silicon N_A at T = 300 K.

Now, initially Fermi-Dirac integral $\mathfrak{F}(\eta_c)$ can be calculated from (1.14) and then the parameter $\eta_c = \frac{E_F - E_c}{kT}$ can be calculated from a set of listed values in [1] and the one which the author of this book to be in the precise form is given below:

$$\eta_c = \frac{ln(\mathfrak{F}(\eta_c))}{1 - \mathfrak{F}(\eta_c)^2} + \frac{\left(3\sqrt{\pi}\frac{\mathfrak{F}(\eta_c)}{4}\right)^{2/3}}{1 + \left[0.24 + 1.08\left(3\sqrt{\pi}\frac{\mathfrak{F}(\eta_c)}{4}\right)^{2/3}\right]^{-2}} \qquad (1.18)$$

Now resubstituting this η_c in the set of Eqs. (1.11) to (1.13), an exact numerical relationship between η_c and Fermi-Dirac integral $\mathfrak{F}(\eta_c)$ can be obtained. This numerically exact $\mathfrak{F}(\eta_c)$ value will be within 0.5 % error margin applied to ionized free carrier concentration based $\mathfrak{F}(\eta_c)$ extraction from (1.14). In Table 1.4, the author shows the value of N_D^+, calculated $\mathfrak{F}(\eta_c)$ from (1.14) and η_c from (1.18). Figure 1.4 shows the trend of this $\mathfrak{F}(\eta_c)$ with respect to η_c.

Next, the free ionized carrier concentration related Fermi-Dirac integral $\mathfrak{F}(\eta_c)$ as listed in Table 1.4 and corresponding η_c values generating the numerically precise $\mathfrak{F}(\eta_c)$ [numerical] using Eqs. (1.11)–(1.13), these three parameters are shown in Table 1.5. Figure 1.5 shows

Table 1.4 The ionized donor density, Fermi-Dirac integral $\mathfrak{F}(\eta_c)$ and η_c for n-type silicon with phosphorous doping in n-type silicon at T = 300 K without band non parabolicity effect

$N_D{}^+$ (/cm^3)	$\mathfrak{F}(\eta_c)$	η_c
2.033×10^{17}	6.294×10^{-3}	-5.065
1.3×10^{18}	0.04025	-3.199
4.535×10^{18}	0.1404	-1.917
1.012×10^{19}	0.3133	-1.056
1.6173×10^{19}	0.5006	-0.519
2.129×10^{19}	0.6591	-0.1864
2.5063×10^{19}	0.7759	0.019
2.7605×10^{19}	0.8546	0.1437
2.923×10^{19}	0.905	0.2189
3.024×10^{19}	0.9362	0.2629

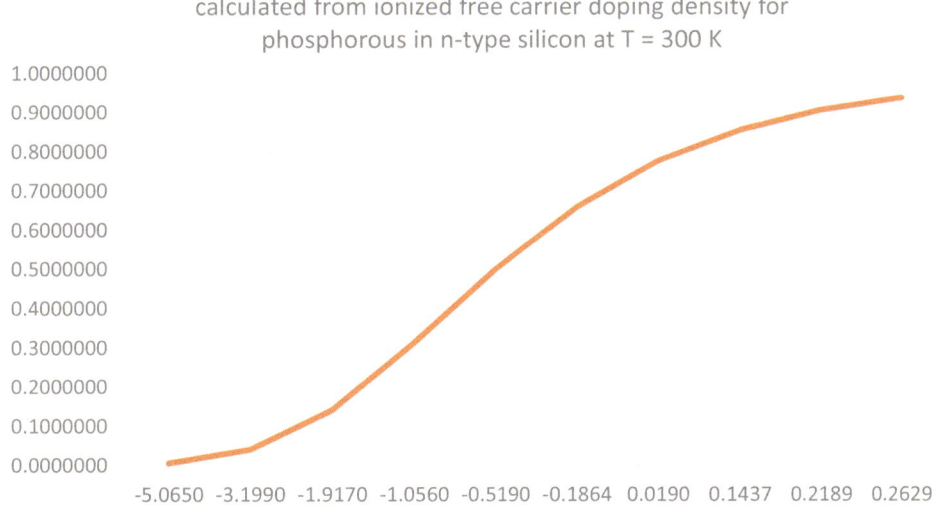

Y axis: Fermi-Dirac Integral $\Im(\eta c)$, X-axis: η_c, calculated from ionized free carrier doping density for phosphorous in n-type silicon at T = 300 K

Fig. 1.4 The Fermi-Dirac integral $\mathfrak{F}(\eta_c)$ as a function η_c for phosphorous doped n-type silicon at T = 300 K without considering band non parabolicity effect. As E_F goes above E_C, $\eta_c > 0$ and $\mathfrak{F}(\eta_c)$ increases with more gradual rate

the plot of $\mathfrak{F}(\eta_c)$ with relation to $\mathfrak{F}(\eta_c)$ ^{numerical} for the same set of η_c calculated using (1.18) for the ionized free carrier density related $\mathfrak{F}(\eta_c)$ values from (1.14).

Table 1.6 shows the deviation of Fermi-Dirac integral $\mathfrak{F}(\eta_c)$ from Eq. (1.14) from Maxwell-Boltzmann distribution based exponential function exp (η_c) by the reduction factor exp (η_c)/ $\mathfrak{F}(\eta_c)$. Figure 1.6 shows the corresponding plot.

Table 1.5 $\mathfrak{F}(\eta_c)$ and $\mathfrak{F}(\eta_c)$ numerical for the same set of η_c values for phosphorous doped n-type silicon at T = 300 K without considering band non parabolicity effect

$\mathfrak{F}(\eta_c)$ numerical	$\mathfrak{F}(\eta_c)$	η_c
6.3089×10^{-3}	6.294×10^{-3}	-5.065
0.040378	0.04025	-3.199
0.13999	0.1404	-1.917
0.31168	0.3133	-1.056
0.49955	0.5006	-0.519
0.65909	0.6591	-0.1864
0.77698	0.7759	0.019
0.85636	0.8546	0.1437
0.9072	0.905	0.2189
0.93803	0.9362	0.2629

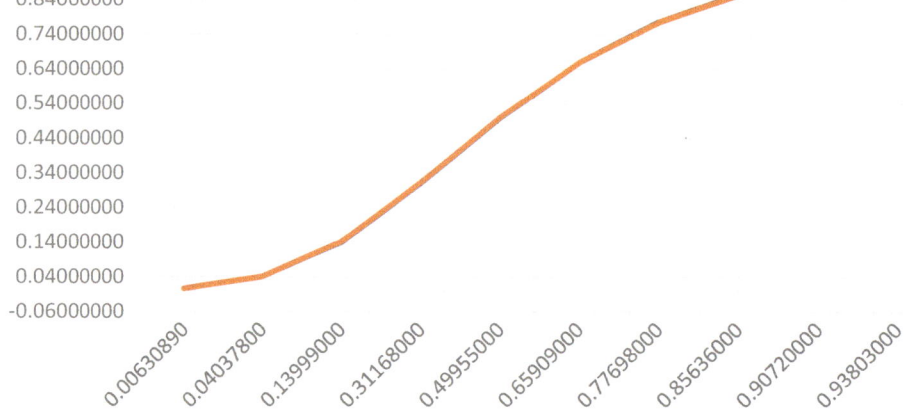

X-axis: Fermi-Dirac integral $\mathfrak{I}(\eta_c)^{numerical}$ (calculated from numerical equations relating $\mathfrak{I}(\eta_c)$ and η_c), Y-axis: Fermi-Dirac Integral $\mathfrak{I}(\eta_c)$ calculated from ionized free carrier density, both for same η_c valueset

Fig. 1.5 Fermi-Dirac integrals $\mathfrak{F}(\eta_c)$ with respect to $\mathfrak{F}(\eta_c)$ numerical for the same set of η_c values of Table 1.5. The trend shows fast rising slope to almost linear and then for more degenerate doping values, slope increases less gradually that means the prediction between numerically computed Fermi-Dirac integral and ionized free carrier concentration Fermi-Dirac integral (without band non parabolicity effect) are more accurate in values as doping goes from high non degenerate to degenerate values

Table 1.6 Reduction factor $\mathfrak{F}(\eta_c)$/exp (η_c) for $\mathfrak{F}(\eta_c)$ calculated from (1.14) and exp (η_c) calculated from same sets of η_c values from (1.18) after (1.14) is used

η_c	$\mathfrak{F}(\eta_c)$	exp (η_c)	$\mathfrak{F}(\eta_c)$/exp (η_c)
−5.065	6.294×10^{-3}	6.314×10^{-3}	0.9962
−3.199	0.04025	0.04080	0.9865
−1.917	0.1404	0.1470	0.9551
−1.056	0.3133	0.3478	0.9000
−0.519	0.5006	0.5951	0.8412
−0.1864	0.6591	0.8299	0.7941
0.019	0.7759	1.0192	0.7613
0.1437	0.8546	1.1545	0.7402
0.2189	0.905	1.2447	0.7271
0.2629	0.9362	1.3007	0.7198

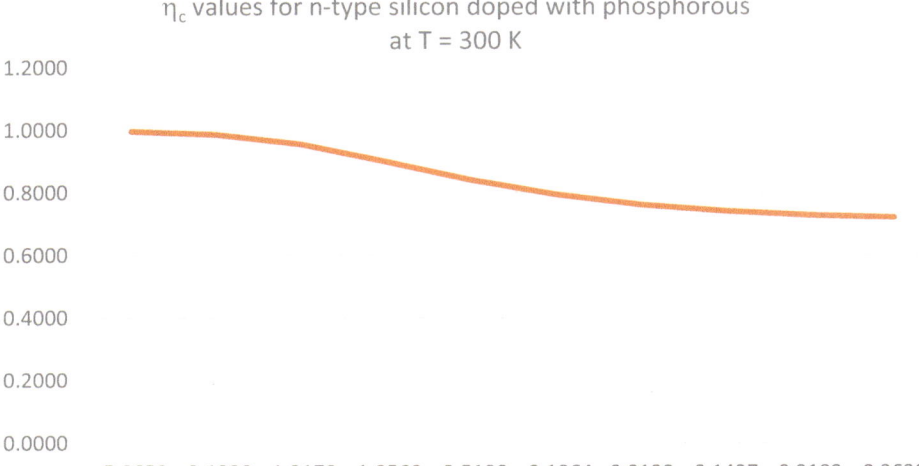

The reduction factor $\Im(\eta_c)$/ exp (η_c) for same set of η_c values for n-type silicon doped with phosphorous at T = 300 K

1.2000									
1.0000									
0.8000									
0.6000									
0.4000									
0.2000									
0.0000									
−5.0650	−3.1990	−1.9170	−1.0560	−0.5190	−0.1864	0.0190	0.1437	0.2189	0.2629

Fig. 1.6 The deviation or reduction factor $\mathfrak{F}(\eta_c)$/exp (η_c) showing with actual ionized dopant related $\mathfrak{F}(\eta_c)$ and exp (η_c) calculations from extracted η_c values from (1.18) using (1.14). The curve decreases more rapidly towards high non degenerate to degenerate doping region of n-type silicon with phosphorous doping at T = 300 K without band non parabolicity effect

We can now proceed with deriving Table 1.7, Fig. 1.7, Table 1.8, Fig. 1.8, Table 1.9 and Fig. 1.9 using the boron doped p-type silicon at T = 300 K without considering band non parabolicity effect. For ionized acceptor dopant concentration $p_o = N_A^-$ based Fermi-Dirac computation and corresponding $\eta_v = \frac{E_v - E_F}{kT}$ extractions, we modify Eqs. (1.14) and (1.18) as:

Table 1.7 The ionized acceptor density, Fermi-Dirac integral $\mathfrak{F}(\eta_v)$ and η_v for p-type silicon with boron doping in p-type silicon at T = 300 K without band non parabolicity effect

$N_A{}^-$ (/cm^3)	$\mathfrak{F}(\eta_v)$	η_v
1.845×10^{17}	0.010084	−4.5926
1.1125×10^{18}	0.060792	−2.7802
4.2089×10^{18}	0.22999	−1.39405
1.0185×10^{19}	0.55654	−0.3927
1.6366×10^{19}	0.89434	0.2031
2.1502×10^{19}	1.175246	0.57465
2.5270×10^{19}	1.3808	0.80658
2.7805×10^{19}	1.5194	0.9494
2.9425×10^{19}	1.60792	1.03596
3.0436×10^{19}	1.66317	1.0883

$$p_o = N_v\mathfrak{F}(\eta_v) \tag{1.19}$$

$$\eta_v = \frac{ln(\mathfrak{F}(\eta_v))}{1 - \mathfrak{F}(\eta_v)^2} + \frac{\left(3\sqrt{\pi}\frac{\mathfrak{F}(\eta_v)}{4}\right)^{2/3}}{1 + \left[0.24 + 1.08\left(3\sqrt{\pi}\frac{\mathfrak{F}(\eta_v)}{4}\right)^{2/3}\right]^{-2}} \tag{1.20}$$

Now this η_v can be used to generated modify equations from (1.11) to (1.13) to calculate numerically precise Fermi-Dirac integral $\mathfrak{F}(\eta_v)$ for p-type silicon with the precise DOS effective mass of hole in silicon at T = 300 K being $m_p = 0.81\ m_o$ [1] where m_o is free electron mass and also considering band non parabolicity effect, which will be discussed in a subsequent Chapter, is not included in these derivations. We list these modified equations for p-type silicon:

$$v(\eta_v) = \eta_v^4 + 50 + 33.6\eta_v\left\{1 - 0.68\exp\left[-0.17(\eta_v + 1)^2\right]\right\} \tag{1.21}$$

$$\varphi(\eta_c) = \frac{3\sqrt{\pi}}{4v(\eta_v)^{3/8}} \tag{1.22}$$

$$\mathfrak{F}(\eta_v) = \left[e^{-\eta_v} + \varphi(\eta_v)\right]^{-1} \tag{1.23}$$

We now discuss some of the salient observations as emanant from the Fermi energy level E_F being significantly higher than E_C, minima of conduction band, with degenerate doping using phosphorous donor doping in n-type silicon at T = 300 K and for the case of the p-type silicon with high degenerate doping levels of boron acceptor doping at T = 300 K. Taking n-type silicon at T = 300 K band structure, as E_F moves significantly higher than E_C, all the energy levels below E_F up to E_C minimum conduction band energy levels will be filled by a greater portion of electrons as free carriers. Taking $N_D{}^+$

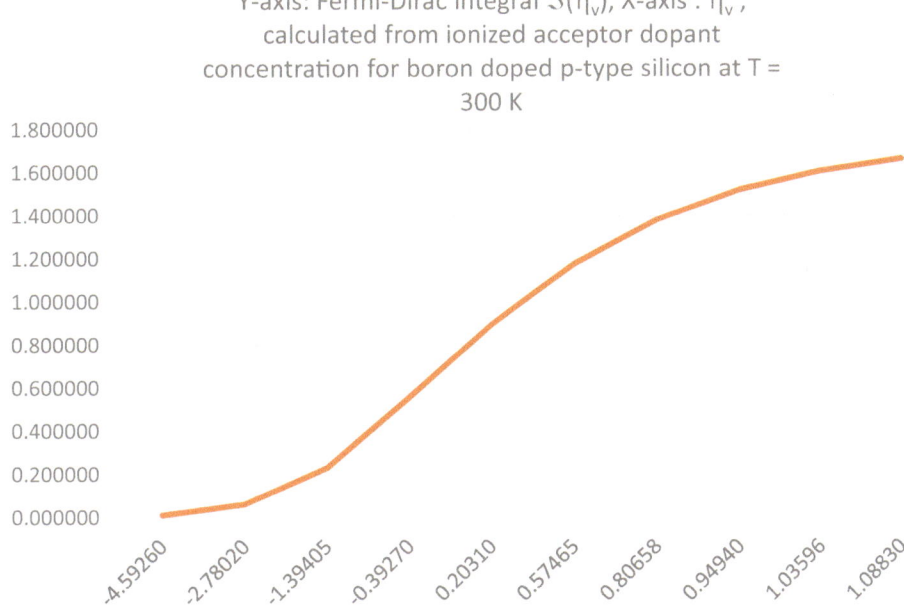

Y-axis: Fermi-Dirac integral $\mathfrak{I}(\eta_v)$, X-axis : η_v , calculated from ionized acceptor dopant concentration for boron doped p-type silicon at T = 300 K

Fig. 1.7 The Fermi-Dirac integral $\mathfrak{F}(\eta_v)$ with respect to η_v calculated for boron doped p-type silicon at T = 300 K without considering band non parabolicity effect. The degeneracy effect and reduced dopant ionization in p-type silicon due to its reduced DOS hole effective mass, makes E_F considerably closer to E_v (reference as 0 ev) for medium non degenerate doping and for high non degenerate to degenerate doping, E_F is considerably below than E_v, resting in deeper level energy levels in the valence band down from valence band maxima E_V which is considered 0 eV or reference energy level. Also Fermi-Dirac integral for boron doped p-type silicon at T = 300 K, rises steeply due to reduced DOS effective hole mass computed N_V and also due to higher ionization of acceptor dopants N_A^- as doping goes from close to high non degenerate to degenerate regime

Table 1.8 $\mathfrak{F}(\eta_v)$ and $\mathfrak{F}(\eta_v)$ numerical for the same set of η_v values for boron doped p-type silicon at T = 300 K without considering band non parabolicity effect

$\mathfrak{F}(v)$ numerical	$\mathfrak{F}(\eta_v)$	η_v
0.010111	0.010084	−4.5926
0.060897	0.060792	−2.7802
0.22878	0.22999	−1.39405
0.55582	0.55654	−0.3927
0.89635	0.89434	0.2031
1.179646	1.175246	0.57465
1.38677	1.3808	0.80658
1.52622	1.5194	0.9494
1.61518	1.60792	1.03596
1.6706	1.66317	1.0883

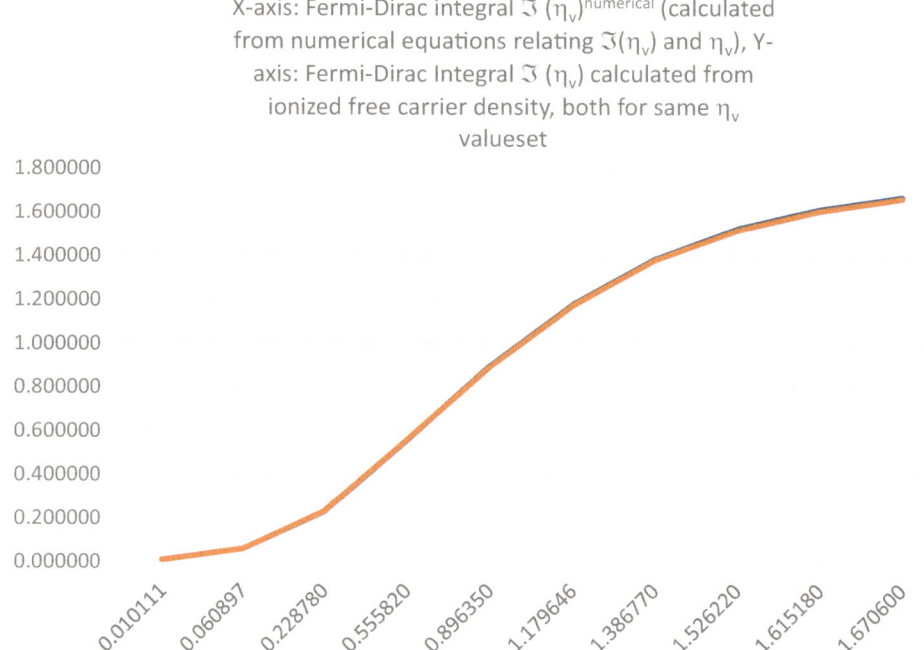

X-axis: Fermi-Dirac integral $\Im(\eta_v)^{numerical}$ (calculated from numerical equations relating $\Im(\eta_v)$ and η_v), Y-axis: Fermi-Dirac Integral $\Im(\eta_v)$ calculated from ionized free carrier density, both for same η_v valueset

Fig. 1.8 Fermi-Dirac integrals $\mathfrak{F}(\eta_v)$ with respect to $\Im(\eta_v)^{numerical}$ for the same set of η_v values of Table 1.8. The trend shows similar like Fig. 1.5, fast rising slope to almost linear and then for more degenerate doping values, slope increases is less gradually that means the prediction between numerically computed Fermi-Dirac integral and ionized free carrier concentration Fermi-Dirac integral (without band non parabolicity effect) are more accurate in values as doping goes from high non degenerate to degenerate values, similar like Fig. 1.5

Table 1.9 Reduction factor $\mathfrak{F}(\eta_v)/\exp(\eta_v)$ for $\mathfrak{F}(\eta_v)$ calculated from (1.19) and exp (η_v) calculated from same sets of η_v values from (1.20) after (1.19) is used

η_v	$\mathfrak{F}(\eta_v)$	$\exp(\eta_v)$	$\mathfrak{F}(\eta_v)/\exp(\eta_v)$
−4.5926	0.010084	0.010126	0.99585
−2.7802	0.060792	0.06203	0.98004
−1.39405	0.22999	0.24807	0.9271
−0.3927	0.55654	0.6752	0.82426
0.2031	0.89434	1.2252	0.72995
0.57465	1.175246	1.77651	0.66155
0.80658	1.3808	2.2402	0.6164
0.9494	1.5194	2.58415	0.58797
1.03596	1.60792	2.8178	0.5706
1.0883	1.66317	2.9692	0.56014

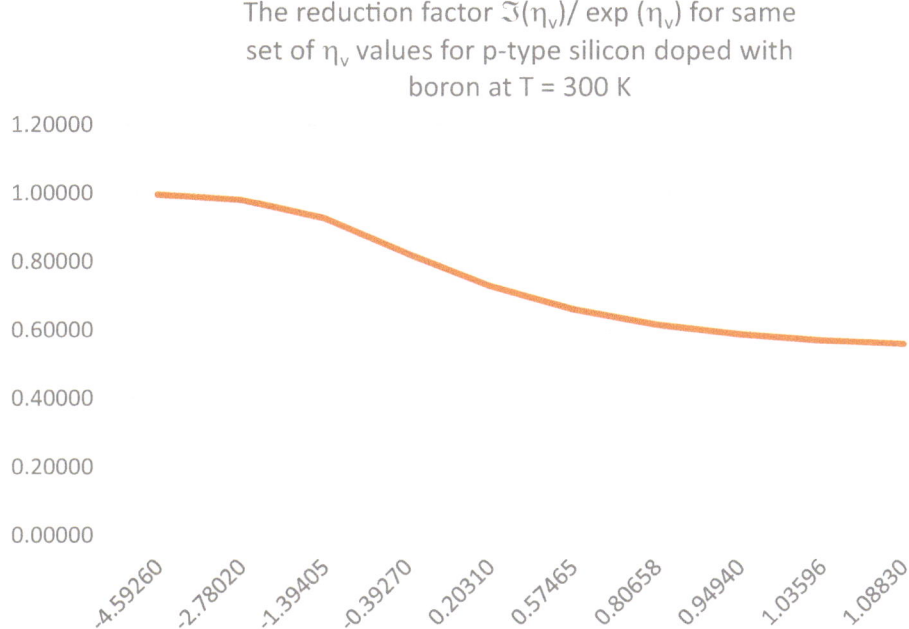

Fig. 1.9 The deviation or reduction factor $\mathfrak{F}(\eta_v)$/exp (η_v) showing with actual ionized acceptor dopant related $\mathfrak{F}(\eta_v)$ and exp (η_v) calculations from extracted η_v values from (1.20) using (1.19). The curve decreases more rapidly with observation of more than decrease of Fig. 1.6 shows, towards high non degenerate to degenerate doping region of p-type silicon with boron doping at T = 300 K without band non parabolicity effect. The reduction factor values decreasing trend is also more sharper at high degenerate doping values compared to Fig. 1.6 decreasing trend which tends to more gradually decreasing at high degenerate doping

$= 3.024 \times 10^{19}$/cm^3, the value of η_c is 0.2629 or E_F is above than E_c by 6.8 meV, so the elevated energy induced momentum increase or kinetic energy increase that will also increase electron drift velocity are such that thermally limited maximum saturation drift velocity related energy balance equilibrium condition is not violated, although these higher energy levels up to 6.8 meV are occupied in greater percentage by electrons, these electrons will not be thermally heated, so carrier heating effect is negligible and degenerate doping induced self heating effect (SHE) as a result of this ionized dopant value range will not occur in majority carrier FET like junctionless n-FET. The drift velocity increase therefore up to 6.8 meV can also aid in junctionless n-FET on current when the device is in accumulation (Gate voltage positive, majority electron carriers accumulate as a channel layer under neath the gate oxide in junctionless n-FET). Now as the donor doping increases to 10^{20}/cm^3 to 10^{21}/cm^3 in the junctionless n-FET with n-substrate boron doped at T = 300 K to form almost ohmic contact at source and drain junction contacts to avoid contact interfacial resistance extending from source contact into the substrate and drain contact into the substrate, η_c values can be greater than 1 and in that case E_F is above than E_C by 25.9 meV and now all the energy level below E_F down to E_C (lowest energy state) are occupied by electrons in greater percentage and the extent of these energy levels occupancy is also in greater range. As a result, electrons at higher energy levels up to E_C to $E_C + 22$ meV can gain exorbitantly high momentum and drift velocity that violates the thermally defined maximum drift velocity for electron at T = 300 K in silicon and the device will be in non equilibrium, carrier heating will take place and self heating effect (SHE) will be intense. So, even though we expect higher drift velocity due to higher kinetic energy of some of these higher energy occupied electrons, the drift velocity will decrease due to carrier heating and on current will be reduced and transconductance of junctionless n-FET will degrade.

Taking the case of p-type silicon with boron doping at T = 300 K, when $N_A^- = 3.0436 \times 10^{19}$/cm^3, the factor $\eta_v = 1.0883$ and E_F being <0 eV is now 28.16 meV below the reference valence band maxima $E_v = 0$ eV. So, even though in n-type silicon due to dopant ionization, the E_F level was not considerably higher than E_C, here in p-type silicon with boron doping at T = 300 K, E_F level is considerably below E_v, which means that all the energy levels up to $E_v = 0$ eV (valence band maxima reference level) are now empty with filling of holes in greater percentage. Holes occupying these distributed energy levels below E_v, gain the required momentum and kinetic energy induced drift velocity close to electron even though holes have lower mobility due to its band structure. Therefore, the carrier heating is taking place with this N_A^- value, a moderate degenerate doping concentration and therefore junctionless p-FET will already face self heating effect (SHE) at this moderate degenerate doping concentration and consequently drift velocity decrease due to carrier heating, on-current decrease compared to n junctionless FET with same substrate doping density 3.07×10^{19}/cm^3 for both n and p junctionless FET. At degenerate doping levels from 10^{20}/cm^3 to 10^{21}/cm^3, p-junctionless FET will suffer from carrier heating and SHE effects more intensely than their n-FET counterpart as the E_F level will be significantly below than E_v in p-type boron doped silicon at T = 300 K than the case of phosphorous doped n-type silicon E_F being above than E_c at T = 300 K. How to reduce this SHE effect when the substrate doping for both n-type and p-type junctionless FET has to be increased degenerately to 10^{21}/cm^3? One way is choosing the substrate doping close to 10^{20} /cm^3 in such a way that interfacial resistance at source and drain contacts, will reduce the lateral electric field and also ensuring for the case of n-type silicon, E_F is moderately higher than E_c and for p-type silicon, E_F is moderately below than E_v, p-type silicon substrate concentration in this case has to be lower than n-type silicon substrate density as analyzed above. This two device engineering introductions will ensure that electrons in n junctionless FET and holes in p-junctionless FET are not subject to carrier heating and SHE related velocity and on-current decrease other than the contact interfacial resistance based decrease and also additionally from transconductance distortion or degradation in both FETs.

References

1. Advanced Semiconductor Fundamentals, Robert F. Pierret. Volume VI, Second edition, Pearson Education Inc., 2003.
2. Intrinsic concentration, effective densities of states, and effective mass in silicon, Martin A. Green, Journal of Applied Physics, 67, 2944 (1990), pp. 2944–2954.
3. Physical Model of Incomplete Ionization for Silicon Device Simulation, Andreas Schenk, Pietro P. Altermatt and Bernhard Schmithúsen, 2006 International Conference on Simulation of Semiconductor Processes and Devices, September 2006, pp. 51–54.
4. Semiconductor Physics and Devices Basic Principles, Donald A. Neamen, Fourth Edition, McGraw Hill Companies Inc, 2012.
5. Device Modeling at Cryogenic Temperatures: Effects of Incomplete Ionization, Akin Akturk, Jeffrey Allnutt, Zeynep Dilli, Neil Goldsman and Martin Peckerar, IEEE Transactions on Electron Devices, Vol. 54, Issue 11, November 2007, pp. 2984–2990.
6. Approximation of Fermi-Dirac integrals of different orders used to determine the thermal properties of metals and semiconductors, O. N. Koroleva, A. V. Mazhukin, V. I. Mazhukin and P. V. Breslavskiy, Physics, 2016.
7. Incomplete ionization in a semiconductor and its implications to device modeling, G. Xiao, J. Lee, J. J. Liou and A. Ortiz-Conde, Microelectronics Reliability, 39, 1999, pp. 1299–1303.

Band Non Parabolicity Introduced DOS Effective Electron Mass and Hole Effective Mass in Silicon at T = 300 K Impact on Related Parameters in Chapter 1

2

Preview of the Chap. 1: Due to non parabolicity effect in n-type and p-type silicon with high degenerate doping at T = 300 K, the DOS effective mass of electron and hole in silicon which is considered to be of fixed values even up to high degenerate doping level in n-type and p-type silicon at T = 300 K, increase sharply non-linearly with the increase factor is higher for DOS effective hole mass in silicon at T = 300 K being subject to more intense band non parabolicity effect than n-type silicon due to p-type silicon's different band structure than n-type extrinsic silicon at T = 300 K.

Degeneracy and band non parabolicity will impact all the parameters calculated in Chap. 1 of this book due to the fact that band non parabolicity in degenerate doping regime in silicon at T = 300 K, increases both DOS effective electron and hole masses non linearly and as a result the equations listed in Chap. 1 where N_C, effective density of states in the conduction band increases by $(m_n)^{3/2}$ and N_V, effective density of states in the valence band increases by $(m_p)^{3/2}$, where thee band non parabolicity effect is more intense in p-type silicon than n-type silicon at high degenerate doping at T = 300 K. The usual procedure of calculating DOS effective electron mass for n-type silicon assuming near parabolic band structure of the conduction band is:

$$m_n = \left(6\sqrt{m_l m_t^2} \right)^{2/3} \tag{2.1}$$

where m_l is longitudinal effective electron mass of silicon and m_t is transverse effective electron effective mass of silicon. Reference [1] is a classic book that at least discusses the non parabolicity induced DOS effective electron mass trend or non linear increase with high degenerate doping. M_l and m_t parameters are also given in [2] and [1] but only assuming non degenerate doping at T = 4.2 K for silicon and [1] where m_l is 0.9163

© The Author(s), under exclusive license to Springer Nature Switzerland AG 2025

N. S. Ashraf, *Parameter-Centric Scaled FET Devices*, Synthesis Lectures on Emerging Engineering Technologies, https://doi.org/10.1007/978-3-031-84286-3_2

m_o and m_t is 0.1905 m_o [2] and also shows the trend of m_l and m_t as a function of temperature from cryogenic temperature to higher than 300 K. The author did not find any instance in available references that m_l and m_t trend being calculated as a function of doping concentration of the substrate where the doping goes from non degenerate to high degenerate, where knowing m_l and m_t in this way, (2.1) will give more precise value of m_n where the band non parabolicity effect is more precisely incorporated. Nonetheless, using [1], we can arrive at some observed m_n trend curve discrete data points as a function of doping where band non parabolicity is intrinsically incorporated:

N_D (/cm^3) m_n/m_o

2×10^{17} 1.18 [1–3]

3×10^{18} 1.2 [approximated from Figure in [1]]

1.5×10^{19} 1.3 [approximated from Figure in [1]]

3×10^{19} 1.4 [approximated from Figure in [1]]

5×10^{19} 1.5 [approximated from Figure in [1]]

Using the above information a polynomial equation with second order has been constructed where the constants in the polynomial terms are calculated using N_D 2×10^{17}/cm^3, 1.5×10^{19}/cm^3 and 5×10^{19}/cm^3 and corresponding m_n/m_o values. The resulted polynomial equation that now incorporates closely approximated band non parabolicity effect is:

$$m_n = 1.1784\, m_o + 0.02652\left(\frac{N_D}{3 \times 10^{18}}\right)m_o - 4.333 \times 10^{-4}\left(\frac{N_D}{3 \times 10^{18}}\right)^2 \qquad (2.2)$$

The normalization value 3×10^{18} is /cm^3 unit.

The results are tabulated in Table 2.1 and Fig. 2.1 shows the m_n/m_o as a function of N_D, actual substrate donor doping at T = 300 K as the DOS band structure for n-type silicon is dependent on N_D but not ionized dopant N_D^+.

Band non parabolicity effect is visible for moderate non degenerate doping level 2.18×10^{17}/cm^3 showing the constant 1.18 m_o of m_n as is used for some textbooks for illustration purposes and problem solving and some referenced articles dated back is inaccurate when considering modern density functional theorem (DFT) or full 3-D quantum simulation are performed. even for modern DFT technique that extracts parameter from silicon 3-D band structure of conduction and valence band, there is no mechanism to intrinsically extract the m_n as a function of N_D in these simulations as we can see in Fig. 2.1.

For p-type silicon at T = 300 K, valence band structure is more complicated with light mass hole band, heavy mass hole band and lower in energy peak split-off band. Therefore, at high degenerate doping, band non parabolicity factor calculation is more challenging and here to circumvent that, a ratio or upscaling based m_p/m_o for DOS effective mass of hole in silicon at T = 300 K is defined although reference [1] had no such plot of DOS effective hole mass in silicon as a function of acceptor substrate doping at T = 300 K.

Table 2.1 DOS electron effective mass normalized to free electron mass, m_n/m_o as a function of doping concentration N_D (/cm^3) in n-type silicon at T = 300 K that approximates the band non parabolicity effect close to the precise extraction

N_D (/cm^3)	m_n/m_o
2.18×10^{17}	1.1803
1.61×10^{18}	1.1925
5.34×10^{18}	1.2242
1.1×10^{19}	1.2698
1.69×10^{19}	1.31405
2.19×10^{19}	1.3489
2.56×10^{19}	1.3732
2.81×10^{19}	1.389
2.97×10^{19}	1.399
3.07×10^{19}	1.4044

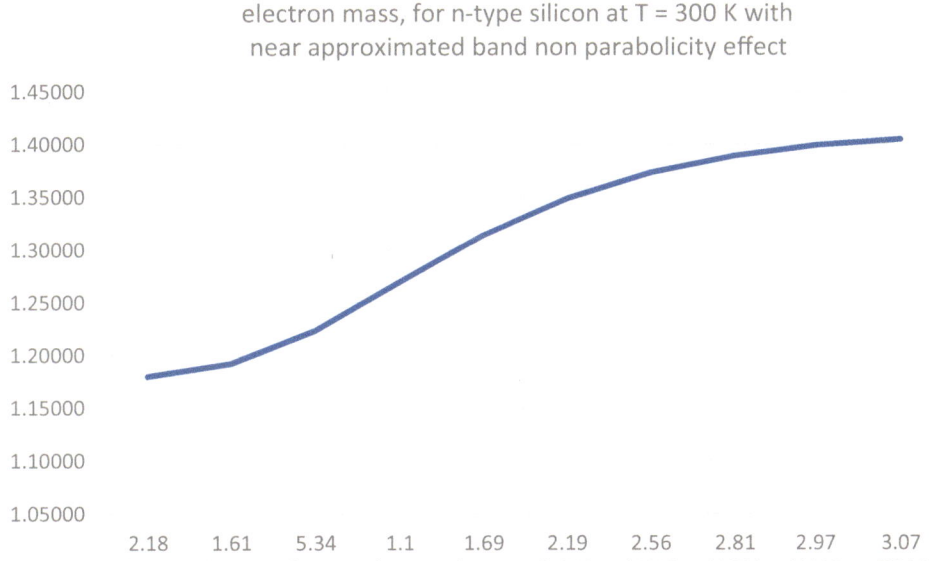

Fig. 2.1 DOS effective electron mass normalized to free electron mass m_n/m_o for n-type silicon at T = 300 K taking the effect of band non parabolicity

Similar like electron effective mass DOS, hole DOS effective mass assuming spherical structure of the valence band shape, the light hole mass or m_{lh}, heavy hole mass m_{hh} and split off band m_{so} are not extracted as a function of doping where band non parabolicity is automatically included in the measured m_{lh}, m_{hh} and m_{so} and references do not have such evidence that semiconductor professionals can compute for their analysis of n-FET and p-FET performance assessment. Here, we include the DOS effective mass m_p as a function of m_{lh} and m_{hh}.

$$m_p = \left(m_{lh}^{3/2} + m_{hh}^{3/2}\right)^{2/3} \tag{2.3}$$

Considering that band non parabolicity related DOS effective hole mass increase in p-type silicon is more than DOS effective electron mass increase in n-type silicon with degenerate doping, the author of this book introduced an approximal viable determination method of m_p as a function of acceptor substrate doping N_A in p-silicon at T = 300 K.

N_A (/cm^3) m_p/m_o

2×10^{17} 0.81 [1–3]

3×10^{18} 0.8505 (Scaling upgrading from previous value 1.05)

1.5×10^{19} 0.8951 (Scaling upgrading from previous value 1.106)

3×10^{19} 0.972 (Scaling upgrading from previous value 1.2)

5×10^{19} 1.053 (Scaling upgrading from previous value 1.3)

A polynomial equation similar like Eq. (2.2) is now constructed for DOS m_p as a function of acceptor substrate doping in p-type silicon at T = 300 K. To evaluate the constant terms, values at $2 \times 10^{17}/\text{cm}^3$, $1.5 \times 10^{19}/\text{cm}^3$ and $5 \times 10^{19}/\text{cm}^3$ are used from above. The normalization value 3×10^{18} is /cm^3 unit.

$$m_h = 0.8088 + 0.01838\left(\frac{N_A}{3 \times 10^{18}}\right) - 2.241 \times 10^{-4}\left(\frac{N_A}{3 \times 10^{18}}\right)^2 \tag{2.4}$$

Table 2.2 and Fig. 2.2 show the non parabolicity effect induced DOS m_p normalized to m_o as a function substrate doping up to degenerate regime in p-type silicon at T = 300 K. As required, the non parabolicity effect will be evident as more intense in p-type silicon for degenerate acceptor doping values.

Even though $N_A = 2.18 \times 10^{17}$ /cm^3 is a non degenerate value, slight non parabolicity effects seen here with $m_p/m_o = 0.8101$. So the research articles published to-date that quoted m_p as $0.81\ m_o$ for p-silicon at T = 300 K for non degenerate doping values (generally considered <10^{18}/cm^3), have done imprecise modeling of p-silicon substrate and n-FET built on p-silicon substrate parameter extraction.

Now, we recompute as precise form the ionized donor dopant concentration for phosphorous in n-silicon at T = 300 K as a function of doping from high non degenerate to degenerate including band non parabolicity effect. In the calculation process, N_c has to substituted by N_c ($m_n = 1.18\ m_o$) $\times ((m_n/m_o)/1.18)^{3/2}$. Table 2.3 and Fig. 2.3 show the ionized dopant percentage for phosphorous doping in n type silicon at T = 300 K for the same substrate doping concentration values used in Table 1.1.

Table 2.2 DOS hole effective mass normalized to free electron mass, m_p / m_o as a function of doping concentration N_A (/cm^3) in p-type silicon at $T = 300$ K that approximates the band non parabolicity effect close to the precise extraction

N_A (/cm^3)	m_p/m_o
2.18×10^{17}	0.8101
1.61×10^{18}	0.8186
5.34×10^{18}	0.8408
1.1×10^{19}	0.87318
1.69×10^{19}	0.90522
2.19×10^{19}	0.93103
2.56×10^{19}	0.94932
2.81×10^{19}	0.9613
2.97×10^{19}	0.9688
3.07×10^{19}	0.97343

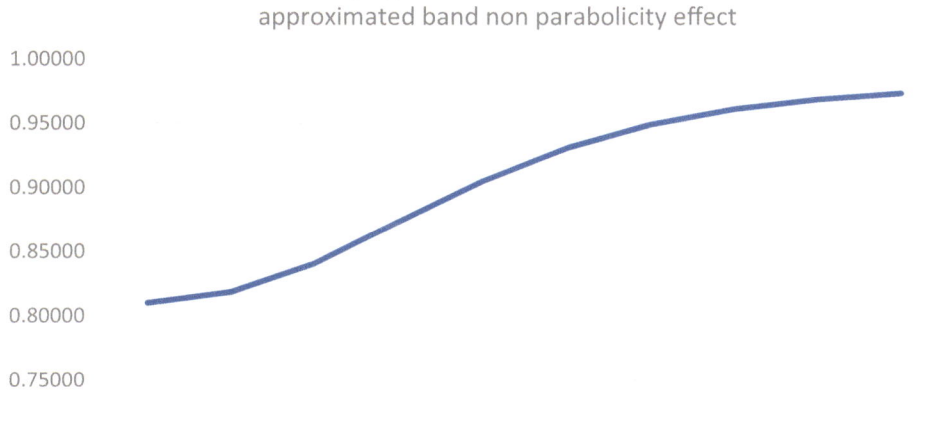

Fig. 2.2 DOS effective hole mass m_p normalized to m_o as a function of substrate doping. Due to small value of m_p, the range of m_p variation percentage considering band non parabolicity is approximately $(0.2/0.81) \times 100 = 24.69\%$ whereas in Fig. 2.2, DOS electron mass m_n variation considering band non parabolicity is $((1.41-1.18)/1.18) \times 100 = 19.49\%$ considering the data extent in the Y-axis of Figs. 2.1 and 2.2 and it clearly shows that Fig. 2.2 has larger variation of DOS hole effective mass in p-type silicon at $T = 300$ K with the incorporation of band non-parabolicity effect

Table 2.3 Ionized donor dopant and its percentage of ionization for phosphorous doped n-type silicon at T = 300 K with inclusion of degeneracy induced band non parabolicity

N_D (/cm^3)	N_D^+ (/cm^3)	N_D^+/N_D (%)
2.18×10^{17}	2.033×10^{17}	93.257
1.61×10^{18}	1.3035×10^{18}	80.963
5.34×10^{18}	4.5635×10^{18}	85.458
1.1×10^{19}	1.0177×10^{19}	92.52
1.69×10^{19}	1.623×10^{19}	96.036
2.19×10^{19}	2.1345×10^{19}	97.466
2.56×10^{19}	2.511×10^{19}	98.09
2.81×10^{19}	2.765×10^{19}	98.4
2.97×10^{19}	2.9269×10^{19}	98.55
3.07×10^{19}	3.0283×10^{19}	98.64

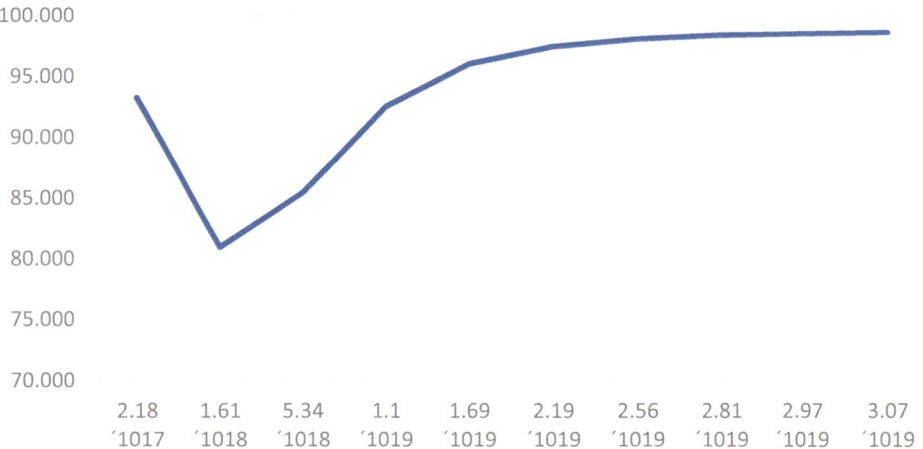

Y-axis ionization percentage, X-axis silicon substrate dopant values (/cm^3) including band non parabolicity effect

Fig. 2.3 Ionization percentage for phosphorous dopant in n-type silicon at T = 300 K including the effect of band non parabolicity effect and the increase of DOS effective electron mass m_n shown in Fig. 2.2 from high non degenerate to degenerate substrate doping values

Initially, with increase of DOS electron effective mass m_n with doping, the ionization percentage variation is more with an increasing trend than Fig. 1.1, but with high degenerate doping, this variation is seen as minute as both Fig. 1.1 and Fig. 2.3 show that when the ionization is close to 100%, variation in N_D^+ and ionization percentage including band non parabolicity effect does not alter these computed values as very conspicuous.

Table 2.4 Actual ionization values and percentage of ionization in p-type silicon at T = 300 K for boron doping with band non parabolicity effect

N_A (/cm^3)	N_A^- (/cm^3)	N_A^-/N_A (%)
2.18×10^{17}	1.8254×10^{17}	83.734
1.61×10^{18}	1.11716×10^{18}	69.39
5.34×10^{18}	4.23539×10^{18}	79.314
1.1×10^{19}	1.02155×10^{19}	92.832
1.69×10^{19}	1.63859×10^{19}	96.958
2.19×10^{19}	2.15169×10^{19}	98.251
2.56×10^{19}	2.52824×10^{19}	98.759
2.81×10^{19}	2.78167×10^{19}	98.992
2.97×10^{19}	2.94354×10^{19}	99.109
3.07×10^{19}	3.0446×10^{19}	99.173

Now, for p-type silicon with boron doping at T = 300 K, in the derivation of ionized acceptor dopants N_A^- (/cm^3), we need to just alter the N_V, the effective density of states in the valence band (/cm^3) by N_V ($m_p = 0.81\ m_o$) $\times ((m_p/m_o)/0.81)^{3/2}$ at T = 300 K. m_p values are derived and tabulated in Table normalized to m_o. Now, we show Table 2.4 for the incomplete ionization percentage for acceptor dopant boron in p-type silicon at T = 300 K with inclusion of more intense band non parabolicity effect. The dip in the ionization percentage is increased from Fig. 1.2 and overall, the ionization is increased to a higher percentage for hole in p-type silicon with boron doping at T = 300 K as the doping of the substrate is made more degenerate value.

Figure 2.4 shows the ionization percentage for boron doped acceptor doping with the same substrate doping values at T = 300 K when band non parabolicity effect is included. We have already seen more variation in m_p/m_o than m_n/m_o as illustrated in Fig. 2.11 and that is the reason that p-type silicon with degenerate boron acceptor doping causes more ionization near 100% than is possible with phosphorous doping in n-type silicon at T = 300 K when for both case, band non parabolicity effect in p-type valence band structure and n-type conduction band structure are numerically incorporated.

Now, important observation as revealed by Table 1.3 and Fig. 1.3 of Chap. 1 are recomputed here in Table 2.5 and Fig. 2.5 with the inclusion of band non parabolicity effect.

Compared to Tables 1.3, Table 2.5 shows that N_A^-/N_D^+ ratio reaches a lower ceiling at high degenerate doping when non parabolicity effect is included. This is due to relative higher percentage of ionized N_A^- and N_D^+ for both p-type and n-type silicon when band non parabolicity effect is correctly incorporated in modeling where n-FET is built on p-type substrate and p-FET is built on n-type substrate. Figure 2.5 shows the N_A^-/N_D^+ ratio with respect to the chosen substrate data set values for either p-type substrate or n-type substrate.

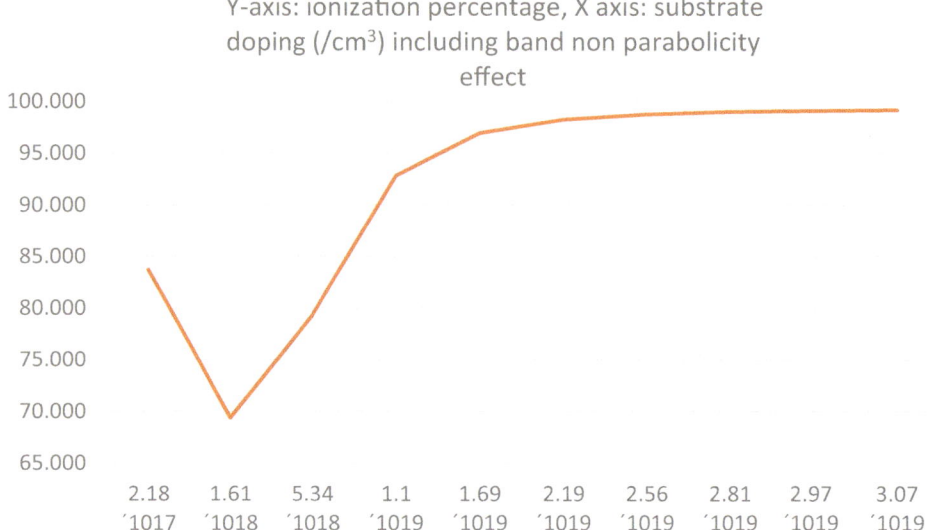

Fig. 2.4 Actual ionization percentage for boron doped acceptor doping in p-type silicon at T = 300 K with the inclusion of band non parabolicity effect due to high degenerate substrate doping. The curve's dip point shows a higher percentage and steeper increase and plateau values at high degenerate doping are also higher compared to Fig. 1.2 where band non parabolicity effect was excluded and constant DOS effective hole mass for silicon at T = 300 K was used

Table 2.5 Actual ionization values and ionization ratio in p-type silicon with respect to n-type silicon at T = 300 K for boron doping (p-type silicon) and phosphorous doping (n-type silicon) with band non parabolicity effect

N_A^- (/cm^3)	N_D^+ (/cm^3)	N_A^-/N_D^+
1.8254×10^{17}	2.033×10^{17}	0.8979
1.11716×10^{18}	1.3035×10^{18}	0.8570
4.23539×10^{18}	4.5635×10^{18}	0.9281
1.02155×10^{19}	1.0177×10^{19}	1.0038
1.63859×10^{19}	1.623×10^{19}	1.0096
2.15169×10^{19}	2.1345×10^{19}	1.0080
2.52824×10^{19}	2.511×10^{19}	1.00686
2.78167×10^{19}	2.765×10^{19}	1.00603
2.94354×10^{19}	2.9269×10^{19}	1.00568
3.0446×10^{19}	3.0283×10^{19}	1.00538

Now, Considering band non parabolicity effect on DOS electron effective mass of silicon for n-type phosphorous dopant at T = 300 K, the Fermi–Dirac integral $\mathfrak{F}(\eta_c)$ and η_c for each doping are now additionally dependent on varying N_C due to band non parabolicity effect and becomes N_C (m_n = 1.18 m_o) × ((m_n/m_o)/1.18)$^{3/2}$ where N_C is

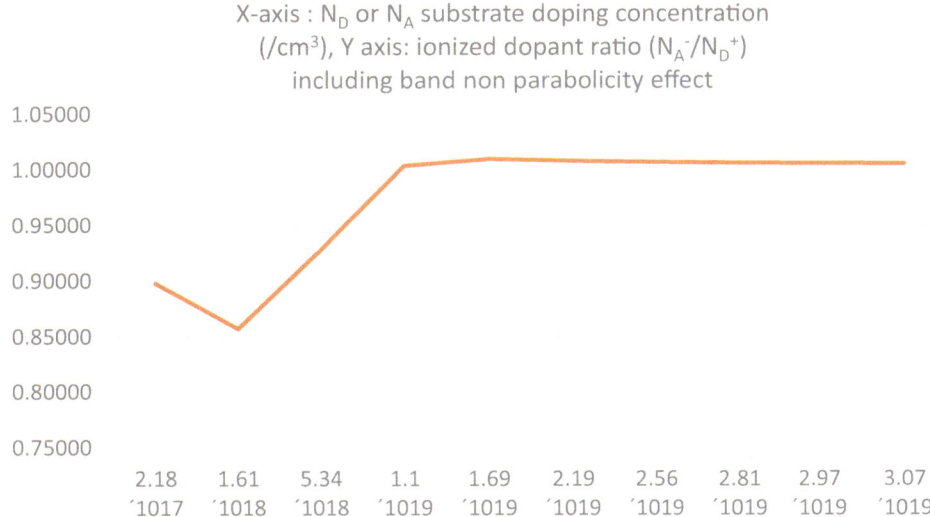

Fig. 2.5 N_A^-/N_D^+ ratio calculated as n-type or p-type silicon substrate doping for silicon at T = 300 K with the inclusion of band non parabolicity. The band non parabolicity with its effect on m_n and m_p, adjusts the N_A^- and N_D^+ and as a result, the ratio ceiling reaches a value near slightly higher than 1 compared to Fig. 1.3, where fixed DOS effective masses for both electron and hole are used to determine ionized dopants with exclusion of band non parabolicity effect

non linearly increasing as the dopant is increased from near degenerate to degenerate doping. What has been computed reveal that $\mathfrak{F}(\eta_c)$ increase ar a lower rate when actual band non parabolicity effect is included in n-type silicon at T = 300 K, than is the case when band non parabolicity effect is excluded or fixed $m_n = 1.18\ m_0$ is used. As a result of which, η_c which changes from negative to positive signed value near ionized donor concentration N_D^+ 2.5063×10^{19}/cm^3 as shown in Tables 1.4 of Chap. 1, now remains negative all the way till ionized donor doping N_D^+ 3.0283×10^{19}/cm^3 (from Table 2.3) as will be shown in the following Table 2.6 for the case of actual band non parabolicity effect included n-type silicon with last maximum percentage ionization computed being 98.64% which is slightly higher than 98.50% as computed in Table 1.1 in Chap. 1 without band non parabolicity effect. More than free carrier variation, N_C increasing towards a wider range from non degenerate to degenerate doping, actually makes η_c negative signed when band non parabolicity effect is considered. But E_D, the donor activation energy now being a band for such high degenerate doping and also with band gap narrowing effect, E_C minima decreasing, the near 100% ionization is sustained and we can safely infer that with band non parabolicity effect included and E_F being lower in position than E_C (with band gap narrowing), the free carriers will reside near the E_C minima energy states and volume inversion will be gate all around (GAA) nanowire junctionless n-FET and for lower gate and drain voltage, upper energy states from E_C which predicted certain

Table 2.6 The ionized donor density, Fermi–Dirac integral $\mathfrak{F}(\eta_c)$ and η_c for n-type silicon with phosphorous doping in n-type silicon at T = 300 K with band non parabolicity effect

$N_D{}^+$ (/cm^3)	$\mathfrak{F}(\eta_c)$	η_c
2.033×10^{17}	6.2917×10^{-3}	-5.0656
1.3035×10^{18}	0.03972	-3.212
4.5635×10^{18}	0.1337	-1.969
1.0177×10^{19}	0.2822	-1.171
1.623×10^{19}	0.4276	-0.7034
2.1345×10^{19}	0.5407	-0.4274
2.511×10^{19}	0.61925	-0.2631
2.765×10^{19}	0.6703	-0.1654
2.9269×10^{19}	0.7019	-0.1081
3.0283×10^{19}	0.7221	-0.07226

level of occupancy by electrons from Table 1.4, where η_c was positive earlier than the most ionized degenerate doping computed from the substrate doping 3.07×10^{19}/cm^3, E_F was then in certain doping values below but close to 3.07×10^{19}/cm^3 substrate doping value, was above E_C (with band gap narrowing effect included) and not only the lowest states close to E_C but certain higher level states were also possible to be occupied by electrons when η_c is moderately fractionally positive, so without band non parabolicity effect, modeling in this way for GAA nanowire junctionless n-FET, not only volume inversion but upper level states in the conduction band are also possible to be occupied by electrons as gate voltage and drain voltage are increases, resulting in carrier temperature rise and self heating effect reported in literature, but with band non parabolicity effect and η_c staying negative even for the last highest substrate doping for n-type phosphorous doped silicon at T = 300 K, it is obvious that states with energies or sub bands to be occupied by electrons are less probable to be occupied by electrons as E_F stays close to E_C but below E_C and hence the inversion carriers will be concentrated in states very close to E_C only and therefore, incorporation of band non parabolicity effect reveals GAA nanowire junctionless n-FET will not be affected by self heating effects (SHE). Now, in Table 2.6 we show the $N_D{}^+$, $\mathfrak{F}(\eta_c)$ and η_c with band non parabolicity effect for phosphorous doped n-type silicon at T = 300 K.

Figure 2.6 shows how $\mathfrak{F}(\eta_c)$ values are decreased with band non parabolicity effect included for phosphorous doped n-type silicon at T = 300 K with effect on η_c remains negatively signed till the last highest value of ionized degenerate carrier density computed and quoted showing that ionization is close to 100% but E_F can be lower in value than E_C when proper band non parabolicity effect due to degeneracy is incorporated.

Now, in Table 2.7 we show the $\mathfrak{F}(\eta_c)$ and $\mathfrak{F}(\eta_c)$ ^{numerical} as calculated from numerical Eqs. (1.11) to (1.13) but now η_c is computed from band non parabolicity effect on DOS electron effective mass m_n as shown in Table 2.6.

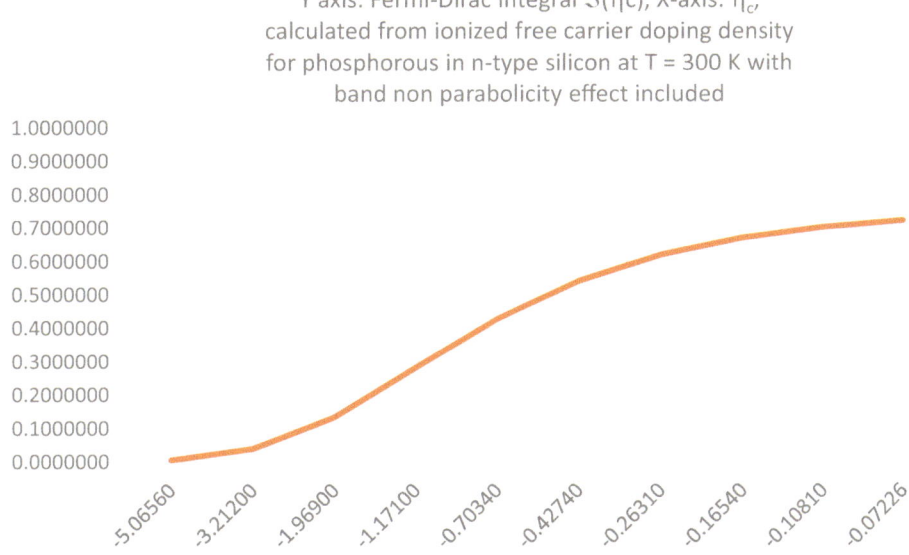

Y axis: Fermi-Dirac Integral $\mathfrak{I}(\eta c)$, X-axis: η_c, calculated from ionized free carrier doping density for phosphorous in n-type silicon at T = 300 K with band non parabolicity effect included

Fig. 2.6 $\mathfrak{F}(\eta_c)$ and corresponding η_c for phosphorous doped n-type silicon at T = 300 K with band non parabolicity effect is included. For all the ionized donor doping values, Fig. 2.6 shows diminution of $\mathfrak{F}(\eta_c)$ to the extent that η_c for the entire doping ranges are negatively signed in stark difference from Fig. 1.4

Table 2.7 $\mathfrak{F}(\eta_c)$ and $\mathfrak{F}(\eta_c)^{\text{numerical}}$ for the new set of η_c values for phosphorous doped n-type silicon at T = 300 K with considering band non parabolicity effect

$\mathfrak{F}(\eta_c)^{\text{numerical}}$	$\mathfrak{F}(\eta_c)$	η_c
6.305×10^{-3}	6.2917×10^{-3}	−5.0656
0.03986	0.03972	−3.212
0.1332	0.1337	−1.969
0.2808	0.2822	−1.171
0.4262	0.4276	−0.7034
0.5399	0.5407	−0.4274
0.619	0.61925	−0.2631
0.6704	0.6703	−0.1654
0.7022	0.7019	−0.1081
0.7227	0.7221	−0.07226

Figure 2.7 shows the $\mathfrak{F}(\eta_c)$ being correlated with numerically calculated using (11)-(13) listed in Chapter 1, $\mathfrak{F}(\eta_c)$ ^{numerical} for phosphorous doped n-type silicon at T = 300 K. η_c are Table 2.6 calculated values in presence of band non parabolicity effect.

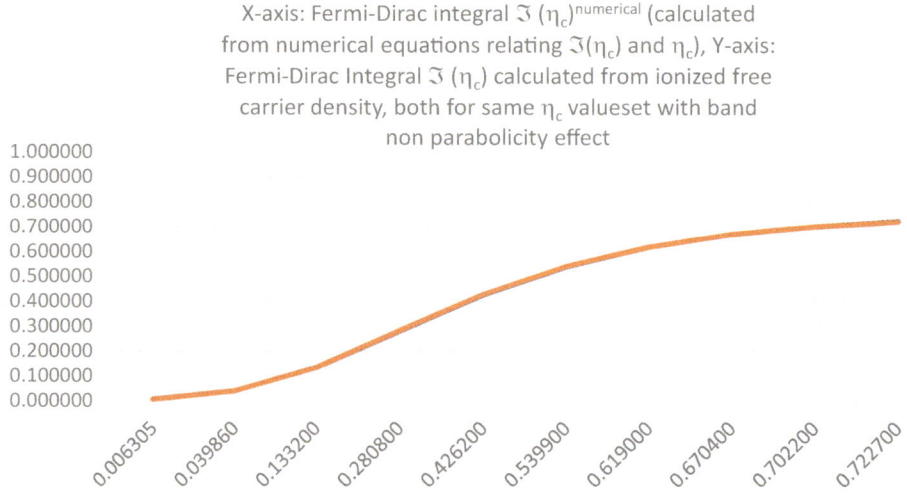

X-axis: Fermi-Dirac integral $\Im(\eta_c)^{numerical}$ (calculated from numerical equations relating $\Im(\eta_c)$ and η_c), Y-axis: Fermi-Dirac Integral $\Im(\eta_c)$ calculated from ionized free carrier density, both for same η_c valueset with band non parabolicity effect

Fig. 2.7 $\mathfrak{F}(\eta_c)$ is correlated in this plot with $\mathfrak{F}(\eta_c)^{numerical}$ for phosphorous doped n-type silicon at T = 300 K. η_c are Table 2.6 calculated values in presence of band non parabolicity effect. Figure 2.7 has similar trend like Fig. 1.5, although the range of values in both axis, are reduced here and device physical based and analytical equations based analysis show here that ionized dopants can be close to 100% but for that Fermi–Dirac integral $\mathfrak{F}(\eta_c)$ may not increase substantially and as a result, η_c may not change sign from negative to positive for very high substrate doping set till the last value chosen here when degeneracy induced band non parabolicity effect is accounted for

Now computation of reduction factor $\mathfrak{F}(\eta_c)/\exp(\eta_c)$ in presence of degeneracy induced band non parabolicity, reveals distinctive difference from Fig. 1.6 in Chap. 1 with parameters computed with fixed DOS electron effective mass m_n in silicon at T = 300 K from non degenerate doping perspective and not assessing the increase of m_n with doping as we have shown in Chapter 2. Due to lower $\mathfrak{F}(\eta_c)$ and η_c values staying negatively signed till the last substrate doping selected for n-type phosphorous doping in silicon at T = 300 K, the ratio values are higher than Fig. 1.6 of Chap. 1 and therefore, proper incorporation of band non parabolicity factor is required as it clearly estimates the relative deviation factor increase than the underestimated values that may have been used in previous modeling results in junctionless n-FET or p-FET built on n-type silicon substrates where the band non parabolicity factor on DOS electron mass increase m_n which has been numerically modeled but not confirmed by further evidences from experimental measurements such as [2] and [3] where these values of effective masses for electron and hole in silicon are measured experimentally at T = 4. 2 K. Table 2.8 shows reduction factor $\mathfrak{F}(\eta_c)/\exp(\eta_c)$ in presence of actual required band non parabolicity in n-type silicon with phosphorous doping at T = 300 K.

Figure 2.8 shows the reduction factor $\mathfrak{F}(\eta_c)/\exp(\eta_c)$ in presence of actual required band non parabolicity in n-type silicon with phosphorous doping at T = 300 K. In contrast

Table 2.8 Reduction factor $\mathfrak{F}(\eta_c)/\exp(\eta_c)$ in presence of actual required band non parabolicity in n-type silicon with phosphorous doping at T = 300 K

η_c	$\mathfrak{F}(\eta_c)$	$\exp(\eta_c)$	$\mathfrak{F}(\eta_c)/\exp(\eta_c)$
−5.0656	6.2917×10^{-3}	6.310×10^{-3}	0.9971
−3.212	0.03972	0.04028	0.9861
−1.969	0.1337	0.1396	0.9577
−1.171	0.2822	0.31006	0.9101
−0.7034	0.4276	0.4949	0.8640
−0.4274	0.5407	0.6522	0.8290
−0.2631	0.61925	0.7687	0.8056
−0.1654	0.6703	0.84755	0.7909
−0.1081	0.7019	0.8975	0.7821
−0.07226	0.7221	0.9303	0.7762

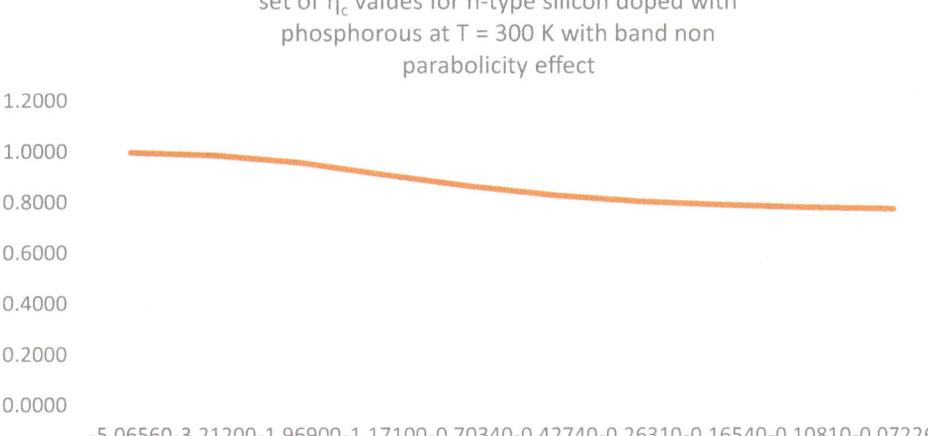

The reduction factor $\Im(\eta_c)/\exp(\eta_c)$ for new same set of η_c values for n-type silicon doped with phosphorous at T = 300 K with band non parabolicity effect

-5.06560 -3.21200 -1.96900 -1.17100 -0.70340 -0.42740 -0.26310 -0.16540 -0.10810 -0.07226

Fig. 2.8 The reduction factor $\mathfrak{F}(\eta_c)/\exp(\eta_c)$ in presence of actual required band non parabolicity in n-type silicon with phosphorous doping at T = 300 K

to Fig. 1.6 the reduction factor values lie in the vicinity of 0.8 for the last substrate donor doping value $3.07 \times 10^{19}/cm^3$ for which $\mathfrak{F}(\eta_c)$ and $\exp(\eta_c)$ both are reduced due to η_c staying negatively signed for all the substrate donor doping values when degeneracy induced band non parabolicity effect is considered.

Now for p-type boron doped silicon at T = 300 K, when more precise band non parabolicity induced DOS effective hole mass increase of m_p is included, near the last substrate degenerating doping chosen here for p-type boron doped silicon at T = 300 K, effective density of states of hole increasing by N_V ($m_p = 0.81\,m_o$) $\times$ ((m_p/m_o)/0.81)$^{3/2}$

from its T = 300 K fixed constant m_p = 0.81 m_0 computed N_V. As a result, $\mathfrak{F}(\eta_v)$ for p-type boron doped silicon near the last value of substrate degenerate doping at T = 300 K, shows a lower values range than what Table 1.7 of Chap. 1 exhibits, but the real difference from degeneracy induced non parabolicity effect in n-type phosphorous doped silicon at T = 300 K shown in Table 2.6 in this chapter is that, η_v starts to becoming positive at much earlier ionized hole carrier concentration. What it means is the Fermi energy E_F is considerably below than top of the valence band E_V, meaning a number of states above E_F with high negative energy level below the reference valence band energy level E_V = 0 eV, will be empty and occupied by holes when the drift field or lateral field is increases such that these energy states are occupied. As a result, at high lateral drift electric field, when high degenerate substrate doping with high hole carrier ionization and more positive η_v are the norm with p-type boron doped silicon at T = 300 K, due to occupancy of these higher negative energy states by holes, their momentum and corresponding drift velocity rise to the level of electron drift velocity limited by thermal electron velocity in silicon at T = 300 K and also by thermal hole velocity in silicon at T = 300 K. The only reason hole drift velocity at high lateral electric field does not rise up to maximum drift velocity of electron at T = 300 K, is owing to higher ionized carrier scattering (hole ionization at T = 300 K is more than electron in silicon for degenerate doping), higher optical phono scattering, higher surface roughness scattering due to more hole ionization as degenerate substrate doping is increased and also due to more carrier heating induced self heating effects (SHE), all of which decrease the mobility at high lateral field for hole as carrier and therefore, the increment of saturation drift velocity of ionized hole as a carrier in junctionless p-FET happens at a lower rate near high lateral field and degenerate doping than the increment noticed for ionized electron as carriers in junctionless n-FET. This is also the reason that both drive or ON current and transconductance in junctionless p-FET with high degenerate substrate doping in presence of band non parabolicity effect and higher drain bias induced lateral drift field, are lower than its counterpart n-type junctionless FET. Table 2.9 shows the ionized acceptor density, Fermi–Dirac integral $\mathfrak{F}(\eta_v)$ and η_v for p-type silicon with boron doping in p-type silicon at T = 300 K with band non parabolicity effect.

Figure 2.9 shows the Fermi–Dirac integral $\mathfrak{F}(\eta_v)$ and corresponding η_v for p-type silicon with boron doping at T = 300 K with band non parabolicity effect.

Table 2.10 shows the $\mathfrak{F}(\eta_v)$ and $\mathfrak{F}(\eta_v)$ ^{numerical} for the η_v of Table 2.9 in presence of band non parabolicity effect.

Figure 2.10 shows the $\mathfrak{F}(\eta_v)$ and numerically computed $\mathfrak{F}(v)^{numerical}$ values for same set of η_v with band non parabolicity effect.

Table 2.11 shows the reduction factor $\mathfrak{F}(\eta_v)$ /exp (η_v) in presence of actual required band non parabolicity in p-type silicon with boron doping at T = 300 K.

As the Table 2.11 shows, reduction factor values $\mathfrak{F}(\eta_v)$/exp (η_v) are higher than Tables 1.9 owing to more gradual increase in exp (η_v) due to less positive η_v values near

Table 2.9 The ionized acceptor density, Fermi–Dirac integral $\mathfrak{F}(\eta_v)$ and η_v for p-type silicon with boron doping in p-type silicon at T = 300 K with band non parabolicity effect

$N_A{}^-$ (/cm^3)	$\mathfrak{F}(\eta_v)$	η_v
1.8254×10^{17}	0.009973	−4.6037
1.11716×10^{18}	0.06009	−2.792
4.23539×10^{18}	0.21884	−1.4476
1.021155×10^{19}	0.49855	−0.5239
1.63859×10^{19}	0.7579	−0.01108
2.15169×10^{19}	0.9541	0.2889
2.52824×10^{19}	1.089	0.468455
2.78167×10^{19}	1.1757	0.57522
2.94354×10^{19}	1.2297	0.6387
3.0446×10^{19}	1.263	0.677

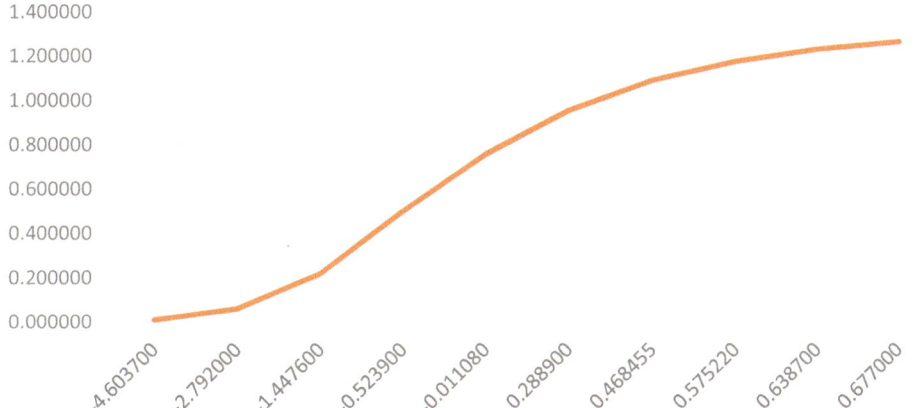

Y-axis: Fermi-Dirac integral $\mathfrak{I}(\eta_v)$, X-axis : η_v, calculated from ionized acceptor dopant concentration for boron doped p-type silicon at T = 300 K with band non parabolicity effect

Fig. 2.9 The Fermi–Dirac integral $\mathfrak{F}(\eta_v)$ and corresponding η_v for p-type silicon with boron doping at T = 300 K with band non parabolicity effect. When correct band non parabolicity effect is included, E_F negative shift from top of the valence band E_V is lower than the case when band non parabolicity effect is excluded (Fig. 1.7, Chap. 1 of this book) near the last recorded degenerate substrate values for acceptor doping. In this range, $\mathfrak{F}(\eta_v)$ values are also reduced from Fig. 1.7, Chap. 1 of this book when band non parabolicity effect is included

Table 2.10 $\mathfrak{F}(\eta_v)$ and $\mathfrak{F}(\eta_v)$ numerical for the same set of η_v values for boron doped p-type silicon at T = 300 K without considering band non parabolicity effect

$\mathfrak{F}(v)^{numerical}$	$\mathfrak{F}(\eta_v)$	η_v
0.01	0.009973	−4.6037
0.0602	0.06009	−2.792
0.2177	0.21884	−1.4476
0.49747	0.49855	−0.5239
0.7587	0.7579	−0.01108
0.9566	0.9541	0.2889
1.0927	1.089	0.468455
1.18	1.1757	0.57522
1.2344	1.2297	0.6387
1.268	1.263	0.677

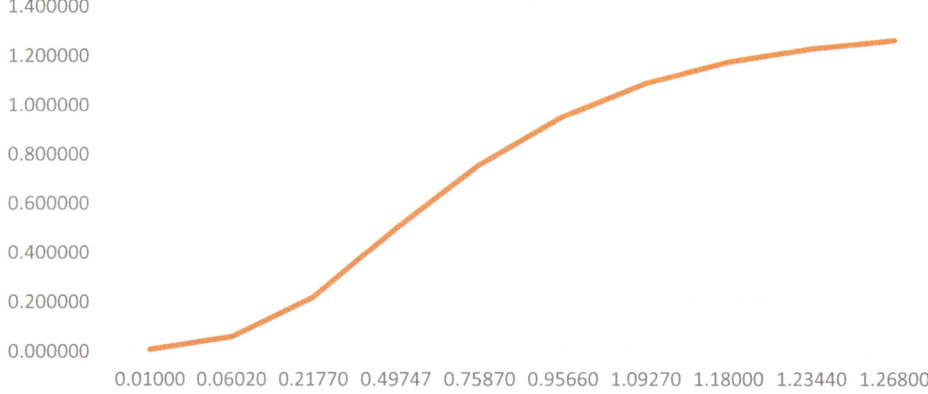

X-axis: Fermi-Dirac integral $\Im(\eta_v)^{numerical}$ (calculated from numerical equations relating $\Im(\eta_v)$ and η_v), Y-axis: Fermi-Dirac Integral $\Im(\eta_v)$ calculated from ionized free carrier density, both for same η_v valueset with band non parabolicity effect

Fig. 2.10 $\mathfrak{F}(\eta_v)$ and numerically computed $\mathfrak{F}(v)^{numerical}$ values for same set of η_v with band non parabolicity effect

the last few degenerate substrate doping values chosen here. It shows that due to degeneracy induced non parabolicity effect, deviation of $\mathfrak{F}(\eta_v)$ values are less than exp (η_v) when compared to more deviation in Tables 1.9.

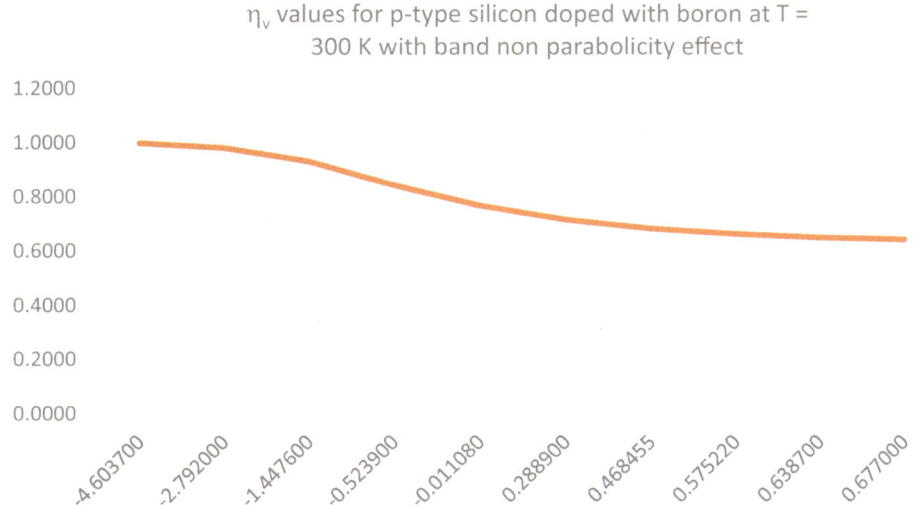

The reduction factor $\Im(\eta_v)/\exp(\eta_v)$ for same set of η_v values for p-type silicon doped with boron at T = 300 K with band non parabolicity effect

Fig. 2.11 The reduction factor $\mathfrak{F}(\eta_v)/\exp(\eta_v)$ for the same set of η_v computed in presence of band non parabolicity effect as a result of degenerate substrate acceptor doping with boron in silicon at T = 300 K. The curve trend and ratio increase compared to values in Fig. 1.9 have already been explained in the discussion of Table 2.11 values trend in the aforementioned paragraph

Table 2.11 The reduction factor $\mathfrak{F}(\eta_v)$ /exp (η_v) in presence of actual required band non parabolicity in p-type silicon with boron doping at T = 300 K.

η_v	$\mathfrak{F}(\eta_v)$	exp (η_v)	$\mathfrak{F}(\eta_v)/\exp(\eta_v)$
−4.6037	0.009973	0.01	0.9973
−2.792	0.06009	0.06129	0.9804
−1.4476	0.21884	0.2351	0.9308
−0.5239	0.49855	0.5922	0.8419
−0.01108	0.7579	0.98898	0.7663
0.2889	0.9541	1.335	0.7147
0.468455	1.089	1.5975	0.6817
0.57522	1.1757	1.7775	0.6614
0.6387	1.2297	1.894	0.6493
0.677	1.263	1.968	0.6418

References

1. Low Temperature Electronics, Physics, Devices, Circuits and Applications, Edmundo A. Gutiérrez-D., M. Jamal Deen and C. Claeys, Academic Press, 2001.
2. Advanced Semiconductor Fundamentals, Robert F. Pierret. Volume VI, Second edition, Pearson Education Inc., 2003.
3. Semiconductor Device Fundamentals, Robert F. Pierret, Addison-Wesley Publication Company Inc, 1996.

Band Non Parabolicity Introduced T = 300 K Key Doping Ionization Related and Carrier Transport Related Parameters Derivation

3

For n-FET fabricated on p-type silicon substrate with boron doping and p-FET fabricated on n-type silicon substrate with phosphorous doping, the drive current of the FET device depends on minority carrier mobility in n and p-FET and saturated drift velocity of minority channel carriers in n and p-FET. Due to surface channel and confinement in n-FET and p-FET, DOS effective masses for electron in n-FET and hole in p-FET cannot be used to calculate inversion channel mobility in n-FET and p-FET which we generally see as doping dependent extraction values. Therefore, to properly determine inversion channel mobility in n and p-FET at T = 300 K in silicon, conductivity minority carrier electron mass for n-FET and conductivity majority carrier hole mass for p-FET are needed to be precisely computed including band non parabolicity effect on valence band for p-substrate where n-FET is built and on conduction band for n-substrate where p-FET is built. Also majority carrier devices such as junctionless transistors and high electron mobility transistors (HEMT), majority carrier conductivity electron effective mass is needed for doping dependent mobility extraction in presence of band non parabolicity on conduction band where the substrate is n-type and majority carrier conductivity hole effective mass is needed for doping dependent mobility extraction in presence of band non parabolicity on valence band where the substrate is p-type. The best way to determine conductivity effective masses for electron and hole in silicon at T = 300 K, considering separate cases for labelling them as majority or minority carriers, doping dependent variation of longitudinal effective mass m_l and transverse effective mass m_t with incorporation of band non parabolicity effect will be crucial for determining conductivity effective masses for electron considering for both majority and minority carriers. On the other hand, doping dependent variation of heavy hole mass m_{hh} and light hole mass m_{lh} with the incorporation of band non parabolicity effect will be crucial for computation of conductivity

N. S. Ashraf, *Parameter-Centric Scaled FET Devices*, Synthesis Lectures on Emerging Engineering Technologies, https://doi.org/10.1007/978-3-031-84286-3_3

effective masses for hole considering for both majority and minority carriers. Gutiérrez [1] is a precise reference which lists the modeling equations of m_l and m_t as a function of temperature considering non-degenerate doping case and related equations shown in [2] can be used to compute electron conductivity effective for silicon as a function of temperature from T = 300 K to lower temperatures. But, this conductivity effective mass for electron is for majority carrier and minority carrier conductivity effective mass for electron from m_l and m_t information, cannot be derived from [2] but since for electron as minority carrier in n-FET, substrate is p-type and in Chap. 2 of this book, it has been shown that band non parabolicity effect is more intense in p-type silicon material, so a relative adjustment procedure based on ratio based increase as shown in derivation of DOS electron and hole mass in presence of band non parabolicity, can be applied in this case also by first determining the electron conductivity effective mass from [2] and then through doping region based ratioed increase, the electron conductivity effective mass as minority carrier can be determined. Gutiérrez-D [1] is also a precise reference which lists modeling equations of m_{hh} and m_{lh} as a function of temperature considering non-degenerate doping and related equations given in [2] can be used to calculate majority carrier hole conductivity mass for silicon as a function of temperature from T = 300 K down to lower temperatures. Again, for minority carrier hole as in p-FET, the conductivity effective mass for hole needs to be ratioed adjustments based on doping regions which are degenerate considering that in p-FET where hole is minority carrier, substrate is n-type and as shown in Chap. 2 of this book, non parabolicity is less intense in n-type silicon due to degeneracy and therefore, adjustment ratioed factor for degenerate doping regime will be different for the case of conductivity effective mass determination of hole as minority carrier.

From [1], the longitudinal and transverse effective mass m_l and m_t for silicon at T = 300 K are m_l = 0.9163 m_o and m_t = 0.2230 m_o and [2] shows the equation for determining conductivity electron effective mass in silicon treating it as majority carrier at T = 300 K assuming non degenerate doping case:

$$\frac{3}{m_{cn}^{maj}} = \frac{2}{m_t} + \frac{1}{m_l} \tag{3.1}$$

Then, m_{cn}^{maj} from the above equation is 0.2982 m_o for electron conductivity effective mass considering non degenerate doping at T = 300 K. Then the following ratio based increase is approximated for conductivity electron effective mass when electron is majority carrier including band non parabolicity effect:

m_{cn}^{maj}/m_o	n-type substrate doping, N_D (/cm^3)
0.2983	10^{17} to 10^{18}
$0.2983 \times 1.05 = 0.3132$	10^{19}
$0.2983 \times 1.08 = 0.3222$	10^{20}

Based on these three regional values, a polynomial equation is constructed where the equation is final form is:

$$\frac{m_{cn}^{maj}}{m_o} = 0.2981 + 4.941 \times 10^{-3}\left(\frac{N_D}{3 \times 10^{18}}\right) - 1.266 \times 10^{-4}\left(\frac{N_D}{3 \times 10^{18}}\right)^2 \quad (3.2)$$

Table 3.1 shows the conductivity electron mass when electron is majority carrier in n-type silicon at T = 300 K with the incorporation of band non parabolicity effect:

Figure 3.1 shows the trend plot of conductivity effective mass normalized to m_o for electron in silicon at T = 300 K when electron is majority carrier including band non parabolicity effect as a function of original substrate doping set in n-type silicon substrate:

Now for majority carrier electron scattering time τ_{cn}^{maj} at T = 300 K for n-type silicon extraction, Drude mobility equation is utilized. Reference [3] is a very good article which gives plot of majority carrier electron mobility as a function of donor doping concentrations up to high degenerate levels $10^{20}/cm^3$, minority carrier electron mobility as a function acceptor substrate doping concentrations up to high degenerate levels $10^{20}/cm^3$, majority carrier hole mobility as a function of acceptor doping concentrations up to $10^{20}/cm^3$ and minority carrier hole mobility as a function of donor doping concentration up to $10^{20}/cm^3$. The Drude mobility equation is mentioned below:

$$\mu_{n=\frac{q\tau_n}{m_{cn}}} \quad (3.3)$$

Equation (3.3) is a generic equation as it can be conditioned to both majority and minority carrier transport related scattering time extractions for both electron and hole in n and p-type silicon or n-FET and p-FET from known conductivity majority and minority effective mass values for electron and hole as carrier and from known majority and minority electron and hole mobility values that come out of modeling based simulation and also from experimental measurements. Using [3], first a "spot and detect" method

	N_D (/cm^3)	m_{cn}^{maj}/m_o
Table 3.1 The conductivity electron mass when electron is majority carrier in n-type silicon at T = 300 K with the incorporation of band non parabolicity effect	2.18×10^{17}	0.29846
	1.61×10^{18}	0.3007
	5.34×10^{18}	0.30649
	1.1×10^{19}	0.314515
	1.69×10^{19}	0.3219
	2.19×10^{19}	0.3274
	2.56×10^{19}	0.33104
	2.81×10^{19}	0.33327
	2.97×10^{19}	0.3346
	3.07×10^{19}	0.3354

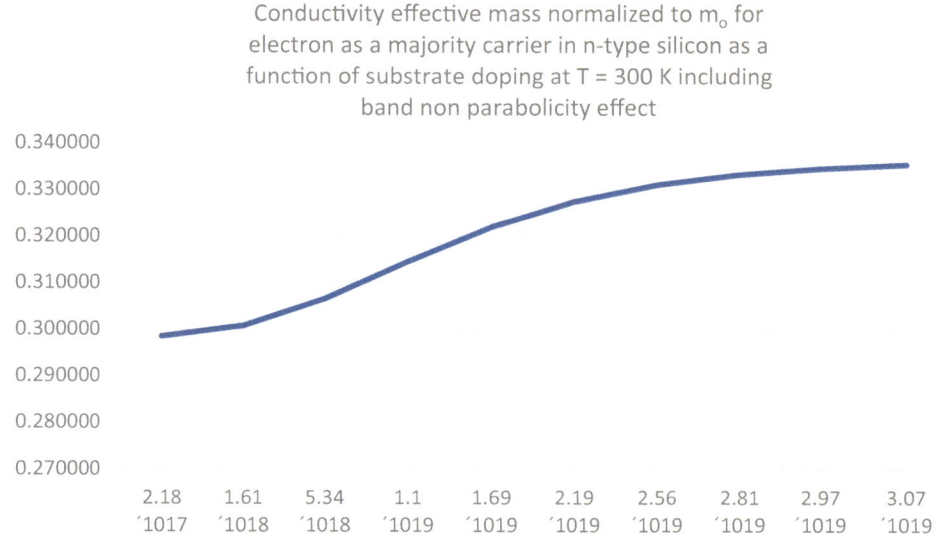

Conductivity effective mass normalized to m_o for electron as a majority carrier in n-type silicon as a function of substrate doping at T = 300 K including band non parabolicity effect

0.340000										
0.330000										
0.320000										
0.310000										
0.300000										
0.290000										
0.280000										
0.270000										
	2.18 '1017	1.61 '1018	5.34 '1018	1.1 '1019	1.69 '1019	2.19 '1019	2.56 '1019	2.81 '1019	2.97 '1019	3.07 '1019

Fig. 3.1 Conductivity effective mass normalized to m_o for electron as a majority carrier in n-type silicon as a function of substrate doping at T = 300 K including band non parabolicity effect. N-type silicon doped with phosphorous has less band non parabolicity effect for degenerate doping concentration as shown in Chap. 2 of this book, as a result m_{cn}^{maj} is proportionately reduced from m_{cn}^{minor} (to be extracted later in this section). Due to this lower m_{cn}^{maj} as will be analyzed later, majority carrier transport related junctionless n-FET and n-HEMT show higher mobility and higher drift velocity

is utilized from a figure to list the majority electron mobility values at certain selected doping concentrations. These values are:

μ_n (cm^2/V–s)	N_D (/cm^3)
730	10^{17}/cm^3
115	10^{19}/cm^3
80	10^{20}/cm^3

Then using Eq. (3.3) and approximated m_{cn}^{maj} values that have been listed previously for N_D, approximated τ_n^{maj} in seconds can be computed. A second order polynomial equation then can be deduced from these reference points and the polynomial equation is listed below:

$$\tau_n^{maj}(s) = 1.39 \times 10^{-14} + 2.179 \times 10^{-14}\left(\frac{N_D}{3 \times 10^{18}}\right)^{-1} - 6.04 \times 10^{-16}\left(\frac{N_D}{3 \times 10^{18}}\right)^{-2}.$$

$$(3.4)$$

Table 3.2 Listing of the majority carrier electron scattering time (s) and its normalized value to 10^{-13} s

N_D (/cm^3)	m_{cn}^{maj}/m_o	τ_n^{maj} (s)	τ_n^{maj} (s)/10^{-13}(s)
2.18×10^{17}	0.29846	1.9942×10^{-13}	1.9942
1.61×10^{18}	0.3007	5.2403×10^{-14}	0.52403
5.34×10^{18}	0.30649	2.5949×10^{-14}	0.25949
1.1×10^{19}	0.314515	1.9798×10^{-14}	0.19798
1.69×10^{19}	0.3219	1.7749×10^{-14}	0.17749
2.19×10^{19}	0.3274	1.6874×10^{-14}	0.16874
2.56×10^{19}	0.33104	1.6445×10^{-14}	0.16445
2.81×10^{19}	0.33327	1.6214×10^{-14}	0.16214
2.97×10^{19}	0.3346	1.6094×10^{-14}	0.16094
3.07×10^{19}	0.3354	1.6023×10^{-14}	0.16023

Here one observation has to be made regarding [3] for different majority an minority carrier based mobility values extractions for electron and hole in silicon at T = 300 K with ionized impurity doping related scattering being considered. As doping is increased to higher from non degenerate to degenerate doping level, threshold voltage of n-FET and p-FET also increase and as a result with same vertical field or gate voltage V_{gs}, gate overdrive or ($V_{gs} - V_T$) decreases where V_T is threshold voltage and it complicates the inversion channel screening effect which modifies additionally the ionized impurity scattering and variation in mobility values extracted. So, at higher doping, these figures need to mention the gate overdrive factor at a relative dopant concentration value for both n-type and p-type silicon and n-FET and p-FET. Also near high degenerate doping levels, the Coulomb or ionized impurity scattering impact is overreached by main mobility decrease that comes out of optical phonon scattering after phonon related peak mobility and then interface or surface roughness scattering, so simply listing the mobility values at high degenerate doping out of increased ionized impurity scattering effect can be unwholesome [3] as additional scattering mechanisms are present when the doping concentration is high and degenerate. Table 3.2 lists the majority carrier electron scattering time (s) and its normalized value to 10^{-13} s as a function of fixed substrate dopant sets. The conductivity electron effective mass values when electron is majority carrier in n-type silicon or FET devices such as junctionless FET and n-HEMT are also mentioned in Table 3.2

Figure 3.2 shows the majority carrier electron scattering time τ_n^{maj} (s) normalized to 10^{-13} s in n-type silicon or majority carrier transport related FET such as junctionless FET and n-HEMT including ionized impurity scattering effect and band non parabolicity effect through the majority carrier electron effective conduction mass m_{cn}^{maj} normalized to m_o.

For minority carrier mobility as a function of doping as modeled in n-FET on p-type silicon substrate and p-FET on n-type silicon substrate at T = 300 K, the effective

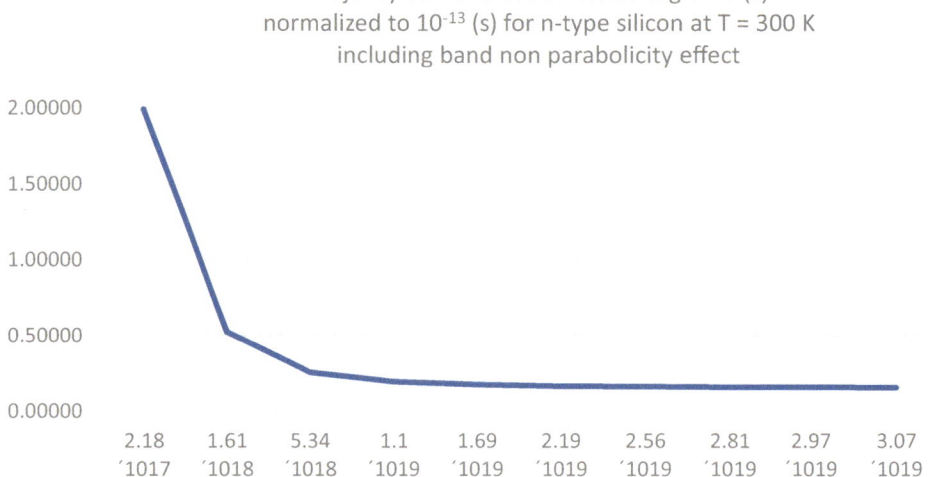

Fig. 3.2 Majority carrier electron scattering time (s) normalized to 10^{-13} s for n-type silicon at T = 300 K including band non parabolicity effect

minority carrier conductivity mass for electron in n-FET and hole in p-FET need to be precisely or approximately analytically precise way computed. Following the procedure of calculating majority carrier electron conductivity mass in n-type silicon at T = 300 K, applying band non parabolicity which becomes more intense for p-type silicon at high degenerate doping making the conductivity electron mass as minority carrier in a p-type acceptor doped substrate, to be higher than conductivity electron mass as majority carrier in native n-type donor doped substrate. Again, considering band non parabolicity effect is negligible for acceptor doping $10^{17}/cm^3$ and $10^{18}/cm^3$, an increasing ratio based increase of electron minority carrier conductivity mass is projected which is shown below:

$m_{cn}{}^{minor}/m_0$	p-type substrate doping, N_A (/cm^3)
0.2983	10^{17} to 10^{18}
$0.2983 \times 1.06 = 0.3162$	10^{19}
$0.2983 \times 1.09 = 0.3251$	10^{20}

Then, a quadratic polynomial coefficient based equation like Eq. (3.2) is formulated using the above data values for determining minority carrier electron conductivity effective mass in p-type silicon at T = 300 K:

$$\frac{m_{cn}^{minor}}{m_o} = 0.2980 + 5.944 \times 10^{-3}\left(\frac{N_A}{3 \times 10^{18}}\right) - 1.54 \times 10^{-4}\left(\frac{N_A}{3 \times 10^{18}}\right)^2 \qquad (3.5)$$

Table 3.3 Listing of the minority carrier electron conductivity effective mass m_{cn}^{minor} normalized to m_o with the substrate doping sets

N_A (/cm^3)	m_{cn}^{minor}/m_o
2.18×10^{17}	0.2984
1.61×10^{18}	0.30114
5.34×10^{18}	0.30809
1.1×10^{19}	0.31772
1.69×10^{19}	0.3286
2.19×10^{19}	0.33318
2.56×10^{19}	0.33751
2.81×10^{19}	0.340164
2.97×10^{19}	0.34175
3.07×10^{19}	0.3427

Table 3.3 lists the minority carrier electron conductivity effective mass m_{cn}^{minor} normalized to m_o with the substrate doping sets considering acceptor doping for p-type substrate with inclusion of band non parabolicity.

Figure 3.3 shows the conductivity effective mass normalized to m_o when electron is minority carrier in transport such as n-FET built on p-type silicon with acceptor boron doping with presence of band non parabolicity effect approximated through ratio based increments of conductivity mass of electron being minority carriers.

Now for minority carrier electron mobility, where in the previous Table and equation, the minority carrier electron conductivity effective mass has been calculated up to degenerate doping considering band non parabolicity, minority carrier scattering time for electron will now be computed for inversion channel mobility related n-FET in p-type silicon material at T = 300 K. By using [4] and "spot and detect" method from the figure of inversion channel mobility of electron as minority carrier in p-type silicon with doping concentration and the m_{cn}^{minor} from the above table at certain doping value using the boundary level values listed, we also list these mobility values at these boundary level values along with doping concentrations:

μ_n (cm^2/V-s)	N_A (/cm^3)
950	10^{17}/cm^3
480	10^{18}/cm^3
200	10^{19}/cm^3
110	10^{20}/cm^3

Now, again a polynomial equation is derived for τ_n^{minor} taking data for 10^{17}/cm^3, 10^{18}/cm^4 and 10^{19}/cm^3. This polynomial equation is listed below:

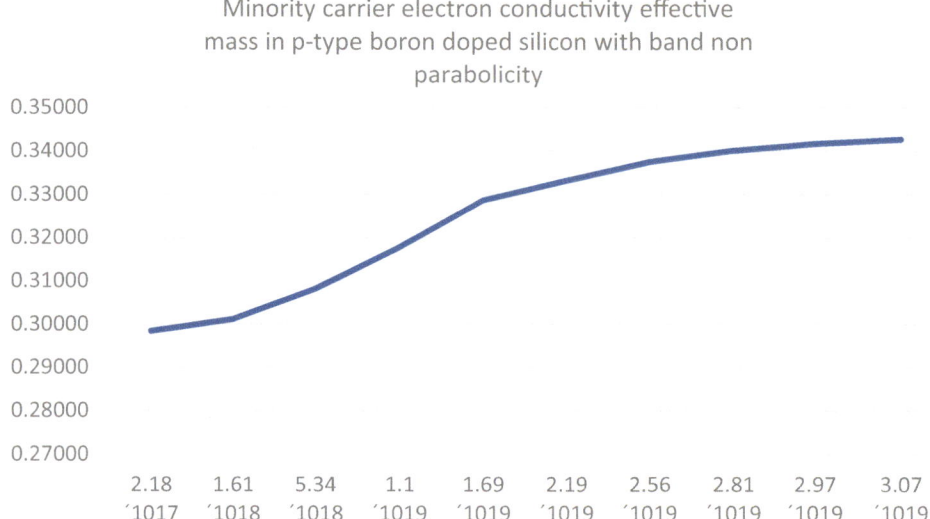

Fig. 3.3 Electron conductivity effective mass normalized to m_o when acting as minority carriers in p-type silicon substrate with boron doping at T = 300 K with the inclusion of band non parabolicity. We can note in this figure, that sharp increase of $m_{cn}{}^{mion}/m_o$ is noted in Fig. 3.3 owing to the reason that band non parabolicity effect is more intense in p-type silicon as the doping becomes increasingly degenerate

$$\tau_n^{minor}(s) = 3.046 \times 10^{-14} + 1.86 \times 10^{-14}\left(\frac{N_A}{3 \times 10^{18}}\right)^{-1}$$

$$- 4.748 \times 10^{-16}\left(\frac{N_A}{3 \times 10^{18}}\right)^{-2} \qquad (3.6)$$

Table 3.4 lists the electron minority carrier scattering time in p-type silicon with band non parabolicity included through extracted mobility values and $m_{cn}{}^{mion}$ conductivity effective mass of electron as carrier.

Figure 3.4 shows the electron scattering time as minority carrier in p-type silicon at T = 300 K and we can observe from Table 3.4, that τ_n^{minor} values are typically larger than τ_n^{maj} values shown in Table 3.2 because, as majority carriers with increase of doping and vertical field from a gate potential in a FET like structure, electrons as majority carrier in junctionless n-FET and n = HEMT, accumulates at the interface making ionized impurity scattering more intense and absence of depletion region underneath the channel for these kind of majority carrier FET and also because of high accumulation with increased doping, the majority electrons face more interface roughness scattering compared to the thin inversion channel for minority carrier electron based FET transport where the effect of inversion layer screening of ionized dopants is not fully extinguished because of the presence thin inversion layer for large vertical field from gate as doping is increased. Also,

Table 3.4 Listing of electron minority carrier scattering time in p-type silicon with band non parabolicity and m_{cn}^{mion} conductivity effective mass of electron as carrier

N_A (/cm^3)	m_{cn}^{minor}/m_0	τ_n^{minor} (s)	τ_n^{minor} (s)/10^{-13}(s)
2.18×10^{17}	0.2984	1.965×10^{-13}	1.965
1.61×10^{18}	0.30114	6.347×10^{-14}	0.6347
5.34×10^{18}	0.30809	4.076×10^{-14}	0.4076
1.1×10^{19}	0.31772	3.5497×10^{-14}	0.35497
1.69×10^{19}	0.3286	3.3747×10^{-14}	0.33747
2.19×10^{19}	0.33318	3.2999×10^{-14}	0.32999
2.56×10^{19}	0.33751	3.2633×10^{-14}	0.32633
2.81×10^{19}	0.340164	3.244×10^{-14}	0.3244
2.97×10^{19}	0.34175	3.2334×10^{-14}	0.32334
3.07×10^{19}	0.3427	3.2273×10^{-14}	0.32273

inversion channel electrons face less surface roughness scattering compared to accumulation channel electrons. The second reason for Table 3.4 showing increased value of τ_n^{minor} is because of m_{cn}^{mionr} values in this case are typically slightly larger than m_{cn}^{maj} values as band non parabolicity effect is larger in p-type acceptor doped substrate with degeneracy. For moderately higher mobility as seen in inversion channel n-FET combined with higher m_{cn}^{mionr} conductivity effective mass for electron as minority carrier, also makes the τ_n^{mionr} values higher through Eq. (3.3).

Table 3.5 shows the relative increase ratio $\tau_{cn}^{minor}/\tau_{cn}^{maj}$, with the ratio saturating at higher doping concentrations, showing the advantage of minority carrier transport based FET devices over majority carrier transport based FET devices. The reasons for this are already explained in the previous paragraph content.

Figure 3.5 shows the relative increment ratio of $\tau_n^{mion}/\tau_n^{maj}$ when assessing benefit of carrier transport based performance of majority carrier FET such as junctionless n-FET and n-HEMT in silicon compared to minority carrier FET such as n-FET fabricated on p-type boron doped silicon at T = 300 K.

For hole transport related conductivity effective mass in silicon at T = 300 K, first we need to construct the conductivity effective mass of hole in p-type silicon from temperature dependent values of light hole mass m_{lh} and heavy hole mass m_{hh}, particularly at T = 300 K. From information taken from [1], these values are $m_{hh} = 0.72\ m_0$ and $m_{lh} = 0.241\ m_0$. Then using the equation provided in [2] which is given below, the conductivity effective mass for hole in silicon at T = 300 K can be computed:

$$m_{cp} = \frac{m_{hh}^{3/2} + m_{lh}^{3/2}}{\sqrt{m_{hh}} + \sqrt{m_{lh}}} \tag{3.7}$$

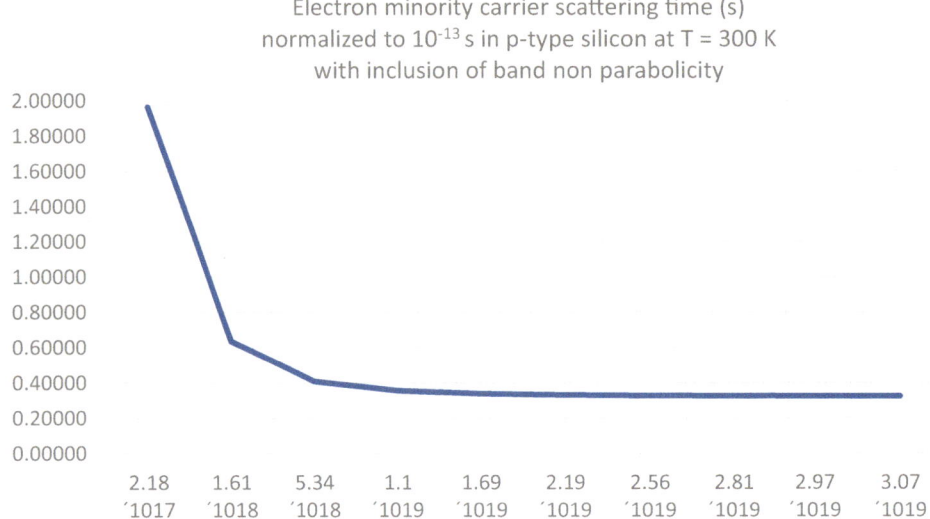

Fig. 3.4 Electron minority carrier scattering time normalized to 10^{-13} s in p-type boron doped silicon substrate at T = 300 K with the inclusion of band non parabolicity

Table 3.5 The relative increase ratio $\tau_{cn}^{minor}/\tau_{cn}^{maj}$

N_A or N_D (/cm^3)	$\tau_n^{minor}/\tau_n^{maj}$
2.18×10^{17}	0.98536
1.61×10^{18}	1.2112
5.34×10^{18}	1.5708
1.1×10^{19}	1.79296
1.69×10^{19}	1.9013
2.19×10^{19}	1.9556
2.56×10^{19}	1.98437
2.81×10^{19}	2.0007
2.97×10^{19}	2.00907
3.07×10^{19}	2.01417

Using (3.7) and the values of m_{lh} and m_{hh} at T = 300 K, the hole conductivity effective mass for non degenerate doping conditions in silicon at T = 300 K is 0.544 m_o. Now, when hole is majority carrier in FET like junctionless p-FET and p-HEMT, the silicon is boron doped p-type and so, with increasing degenerate doping concentration, the non parabolicity effect will be more on m_{cp}^{maj} or majority carrier hole conductivity effective mass in silicon at T = 300 K. Following the ratioed based increment separating the doping

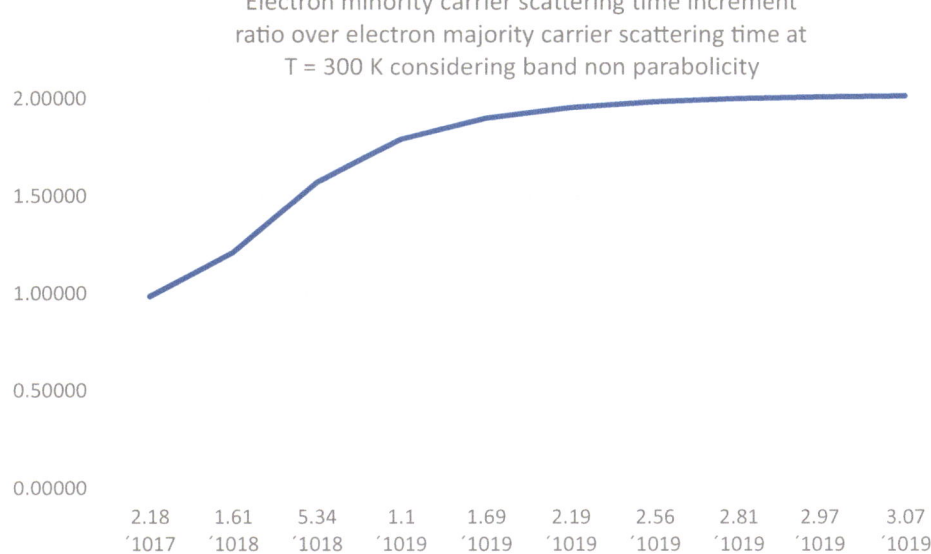

Fig. 3.5 Electron minority carrier scattering time increment ratio with respect to electron majority carrier scattering time. Initially, for non degenerate doping regime, we can see the curve increases faster denoting higher speed and mobility of carriers when these carriers are minority carriers such as n-FET built on p-type silicon substrate. As doping is increased further to degenerate regime and non parabolicity effect is included, the curve starts to saturate as interface roughness scattering and carrier-carrier scattering become dominant in both type substrates and ionized impurity scattering effect or its relative screening by inversion channel in electron minority carrier transport gradually become less important or impactful

regime, we can approximate the band non parabolicity effect on m_{cp}^{maj} by the following way:

m_{cp}^{maj}	N_A
$0.544\ m_o$	$10^{17}/cm^3$
$0.544\ m_o \times 1.07$	$10^{19}/cm^3$
$0.544\ m_o \times 1.1$	$10^{20}/cm^3$

With these boundary equations, a three term polynomial equation is derived for m_{cp}^{maj} which is listed below:

$$\frac{m_{cp}^{maj}}{m_o} = 0.5436 + 0.0126\left(\frac{N_A}{3 \times 10^{18}}\right) - 3.294 \times 10^{-4}\left(\frac{N_A}{3 \times 10^{18}}\right)^2 \tag{3.8}$$

Table 3.6 The values of conductivity effective mass of hole as majority carriers in p-type silicon at T = 300 K with the same basic set of boron doped substrate concentrations

N_A (/cm^3)	m_{cp}^{maj}/m_o
2.18×10^{17}	0.54451
1.61×10^{18}	0.55027
5.34×10^{18}	0.56498
1.1×10^{19}	0.58537
1.69×10^{19}	0.60413
2.19×10^{19}	0.61803
2.56×10^{19}	0.62713
2.81×10^{19}	0.63272
2.97×10^{19}	0.63606
3.07×10^{19}	0.63804

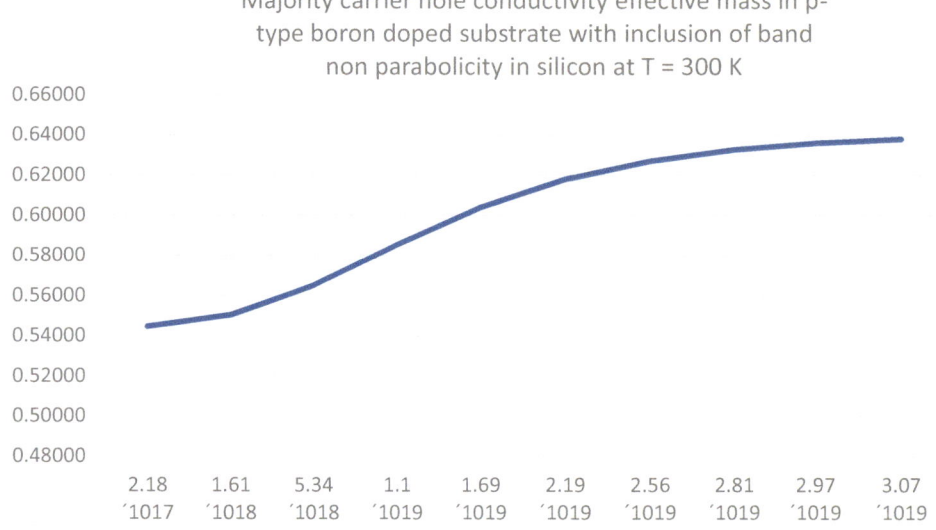

Fig. 3.6 Majority carrier hole conductivity effective mass m_{cp}^{maj} in silicon as a function of p-type acceptor doping concentration at T = 300 K including band non parabolicity effect

Table 3.6 shows the values of conductivity effective mass of hole as majority carriers in p-type silicon at T = 300 K with the same basic set of boron doped substrate concentrations:

Figure 3.6 shows the plot of majority carrier hole conductivity effective mass in silicon at T = 300 K as a function of doping up to degenerate doping and inclusion of band non parabolicity effect.

Now for majority carrier hole scattering time computation in p-type silicon at T = 300 K, first we need to note that in the Drude mobility Eq. (3.3) with τ_n replaced by τ_p and m_{cn} by m_{cp}, it has been noted that m_{cp}^{maj} is proportionally in higher magnitude than m_{cn}^{maj} due to non parabolicity effect incorporation, which changes the scattering time τ_{cp}^{maj} to be higher relative to τ_{cn}^{maj}. Also for majority carrier scattering for both electron and hole in either n-type or p-type substrate, the carriers witness an attractive potential by positively charged donor scatterers for electron and by negatively charged acceptor scatterers for holes. Also positively charged donor dopants have different scattering cross section and Coulomb attractive force surrounding the electrons in transport compared to negatively charged acceptor dopants posing a different scattering cross section and Coulomb attractive force surrounding the holes in transport. We now list the boundary values of majority carrier hole mobility in p-type silicon at T = 300 K as a function of doping up to degenerate regime from figure in [4] by "spot and detect" method.

μ_p (cm^2/V–s)	N_A (/cm^3)
310	10^{17}/cm^3
160	10^{18}/cm^3
70	10^{19}/cm^3

A three term polynomial equation is derived to represent τ_p^{maj} from the above boundary conditions value by substituting the values of μ_p and m_{cp}^{maj} in Eq. (3.3) and computing initial boundary values of τ_p^{maj}. The equation is shown below:

$$\tau_{cp}^{maj}(s) = 2.015 \times 10^{-14} + 1.08 \times 10^{-14}\left(\frac{N_A}{3 \times 10^{18}}\right)^{-1}$$
$$- 2.757 \times 10^{-16}\left(\frac{N_A}{3 \times 10^{18}}\right)^{-2} \tag{3.9}$$

Table 3.7 shows the hole majority carrier scattering time τ_{cp}^{maj} in seconds in p-type silicon with boron doped substrate doping at T = 300 K and the normalized values to 10^{-13} s.

Figure 3.7 shows the plot of hole majority carrier scattering time normalized to 10^{-13} s in p-type boron doped silicon at T = 300 K for substrate doping up to degenerate regime considering band non parabolicity. The impact of band non parabolicity is different for τ_{cp}^{maj} as shown in Table 3.7 from τ_{cn}^{maj} as shown in Table 3.2.

For minority hole transport and mobility in n-type silicon with donor doping at T = 300 K, as we notice that band non parabolicity effect is gradually lesser for degenerately doped n-type silicon substrate compared to p-type acceptor doped substrate. As a result, the conductivity effective mass for hole being minority carriers in n-type phosphorous doped silicon substrate is also proportionally lower than their counterpart in p-type boron doped substrate. This is one reason that minority carrier hole mobility is larger then

Table 3.7 The hole majority carrier scattering time $\tau_{cp}{}^{maj}$ in p-type silicon with boron doped substrate doping at T = 300 K and the normalized values to 10^{-13} s

N_A (/cm^3)	$m_{cp}{}^{maj}/m_o$	$\tau_p{}^{maj}$ (s)	$\tau_p{}^{maj}$ (s)/10^{-13}(s)
2.18×10^{17}	0.54451	1.1656×10^{-13}	1.1656
1.61×10^{18}	0.55027	3.932×10^{-14}	0.3932
5.34×10^{18}	0.56498	2.613×10^{-14}	0.2613
1.1×10^{19}	0.58537	2.3074×10^{-14}	0.23074
1.69×10^{19}	0.60413	2.2058×10^{-14}	0.22058
2.19×10^{19}	0.61803	2.1624×10^{-14}	0.21624
2.56×10^{19}	0.62713	2.1412×10^{-14}	0.21412
2.81×10^{19}	0.63272	2.12999×10^{-14}	0.212999
2.97×10^{19}	0.63606	2.1238×10^{-14}	0.21238
3.07×10^{19}	0.63804	2.1203×10^{-14}	0.21303

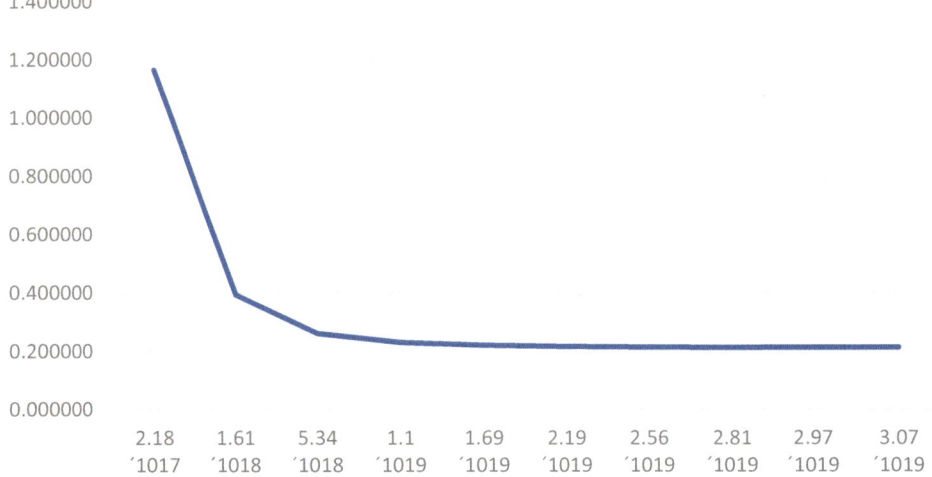

Fig. 3.7 Majority carrier hole scattering time normalized to 10^{-13} s for p-type boron doped silicon at T = 300 K for up to degenerate substrate doping including band non parabolicity. Inclusion of band non parabolicity with doping dependent increase of $m_{cp}{}^{maj}$ changes the scattering time magnitude from the derivation using fixed m_{cp} value deduced from non degenerate doping condition at T = 300 K for p-type silicon

Table 3.8 The $m_{cp}{}^{mior}$ normalized to m_0 as a function of donor substrate doping concentration in silicon at T = 300 K with inclusion of band non parabolicity

N_D (/cm^3)	$m_{cp}{}^{minor}/m_0$
2.18×10^{17}	0.5444
1.61×10^{18}	0.5494
5.34×10^{18}	0.56228
1.1×10^{19}	0.58010
1.69×10^{19}	0.59651
2.19×10^{19}	0.6087
2.56×10^{19}	0.61869
2.81×10^{19}	0.6216
2.97×10^{19}	0.6245
3.07×10^{19}	0.62629

majority carrier hole mobility in silicon at low vertical and lateral field and more explanations for this effect will be covered in the later part of this related discussion. Here we list the boundary values of $m_{cp}{}^{minor}/m_0$ and N_D which are used to derive the three term polynomial equation for hole conductivity effective mass in n-type silicon at T = 300 K when hole carriers are minority.

$m_{cp}{}^{minor}/m_0$	N_D (/cm^3)
0.544	10^{17}
0.544×1.06	10^{19}
0.544×1.09	10^{20}

The three term polynomial equation for $m_{cp}{}^{mion}/m_0$ is given below:

$$\frac{m_{cp}^{minor}}{m_o} = 0.5436 + 0.011\left(\frac{N_D}{3 \times 10^{18}}\right) - 2.853 \times 10^{-4}\left(\frac{N_D}{3 \times 10^{18}}\right)^2 \qquad (3.10)$$

The Table 3.8 shows the $m_{cp}{}^{mior}$ normalized to m_0 as a function of donor substrate doping concentration in silicon at T = 300 K with inclusion of band non parabolicity:

Figure 3.8 shows the plot of minority carrier hole conductivity effective mass $m_{cp}{}^{minor}$ normalized to m_0 in n-type phosphorous doped silicon at T = 300 K with the inclusion of band non parabolicity effect.

For minority carrier hole scattering time determination in n-type silicon substrate at T = 300 K with inclusion of band non parabolicity effect due to degenerate doping concentrations, first three boundary values of minority carrier hole mobility in n-type donor doped silicon at those doping concentration values are selected from reference [4] from provided figure by "spot and detect" approach. These values are listed below:

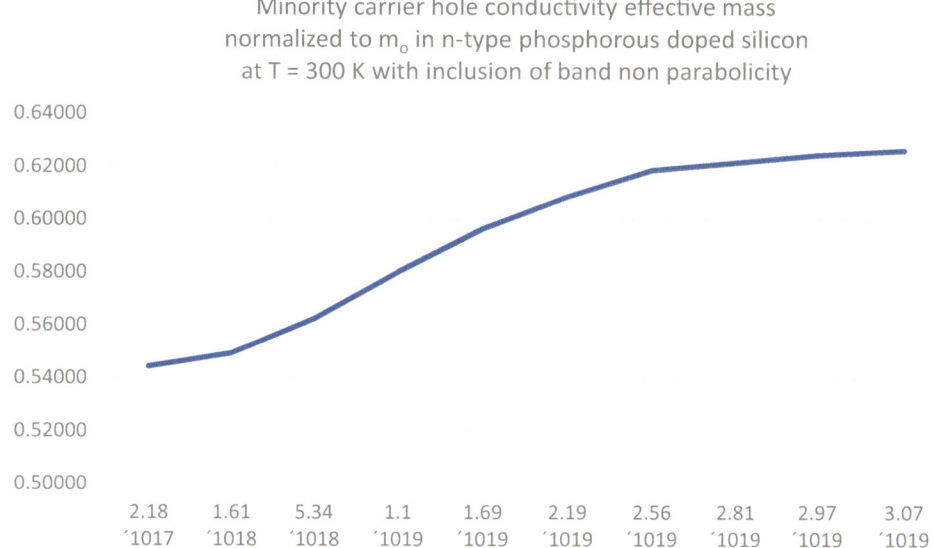

Fig. 3.8 Minority carrier hole conductivity effective mass in silicon at T = 300 K extracted from doping up to degenerate regime in n-type phosphorous doping with the inclusion of band non parabolicity effect

μ_n (cm^2/V–s)	N_D (/cm^3)
440	10^{17}/cm^3
325	10^{18}/cm^3
150	10^{19}/cm^3

Now corresponding three boundary values of minority carrier hole scattering time $\tau_{cp}{}^{minor}$ can be calculated from Eq. (3.3) by substituting μ_p for μ_n and $m_{cp}{}^{minor}$ for m_{cn}. Now a three term polynomial equation is derived for $\tau_{cp}{}^{minor}$ (s) which is shown below:

$$\tau_{cp}^{minor}(s) = 4.303 \times 10^{-14} + 2.12 \times 10^{-14}\left(\frac{N_D}{3 \times 10^{18}}\right)^{-1}$$

$$- 6.03 \times 10^{-16}\left(\frac{N_D}{3 \times 10^{18}}\right)^{-2} \tag{3.11}$$

Table 3.9 The minority conductivity effective mass of hole normalized to m_o and minority hole scattering time normalized to 10^{-13} s with band non parabolicity effect included

N_D (/cm^3)	m_{cp}^{minorj}/m_o	τ_p^{minor} (s)	τ_p^{minor} (s)/10^{-13}(s)
2.18×10^{17}	0.5444	2.20578×10^{-13}	2.20578
1.61×10^{18}	0.5494	8.0439×10^{-14}	0.80439
5.34×10^{18}	0.56228	5.47498×10^{-14}	0.547498
1.1×10^{19}	0.58010	4.8769×10^{-14}	0.48769
1.69×10^{19}	0.59651	4.6774×10^{-14}	0.46774
2.19×10^{19}	0.6087	4.5923×10^{-14}	0.45923
2.56×10^{19}	0.61869	4.5506×10^{-14}	0.45506
2.81×10^{19}	0.6216	4.52865×10^{-14}	0.452865
2.97×10^{19}	0.6245	4.51653×10^{-14}	0.451653
3.07×10^{19}	0.62629	4.50959×10^{-14}	0.450959

Table 3.9 shows the minority carrier hole scattering time τ_{cp}^{minor} in seconds in donor doped n-type silicon at T = 300 K for doping concentrations up to degenerate regime where inclusion of non parabolicity effect has been included from the Drude equation by modified m_{cp}^{mionor} values, although the mobility values are modeling based rather than experimental based [4] and though Hall mobility approach gives the majority carrier mobility where the non parabolicity effect in the degenerate doping concentrations is inherently accounted for, inversion channel mobility by experimental method is very difficult for extraction even with precision method as these values cannot be extracted directly from measurements and some additional numerical parameter extraction based calculations are required.

Figure 3.9 shows the hole minority carrier scattering time normalized to 10^{-13} s in n-type phosphorous doped silicon substrate at T = 300 K.

Comparing Fig. 3.9 with Fig. 3.7, we see that hole minority carrier scattering time is larger than home majority carrier scattering time and this can be interpreted by observation that from Drude mobility equation, m_{cp}^{mionr} is increasing at a progressively slower rate than m_{cp}^{maj} as the band non parabolicity in n-type silicon is less intense than p-type silicon with degenerate doping and also another observation is that minority carrier hole mobility values are also larger than majority carrier hole mobility values which we see in both n-FET on p-silicon and p-FET on n-silicon, a result mostly out of repulsive field of scattering zone and inversion layer screening of ionized impurity dopants when minority carrier mobility transport happens, where as energetic electrons and holes experience attractive force-field within the scattering cross section with the ionized dopants when majority carrier mobility is computed for n-type majority FET devices such as junctionless n-FET and n-HEMT and also for p-type majority FET devices such as junctionless p-FET and p-HEMT. There are physical aspects emanating from interface and surface

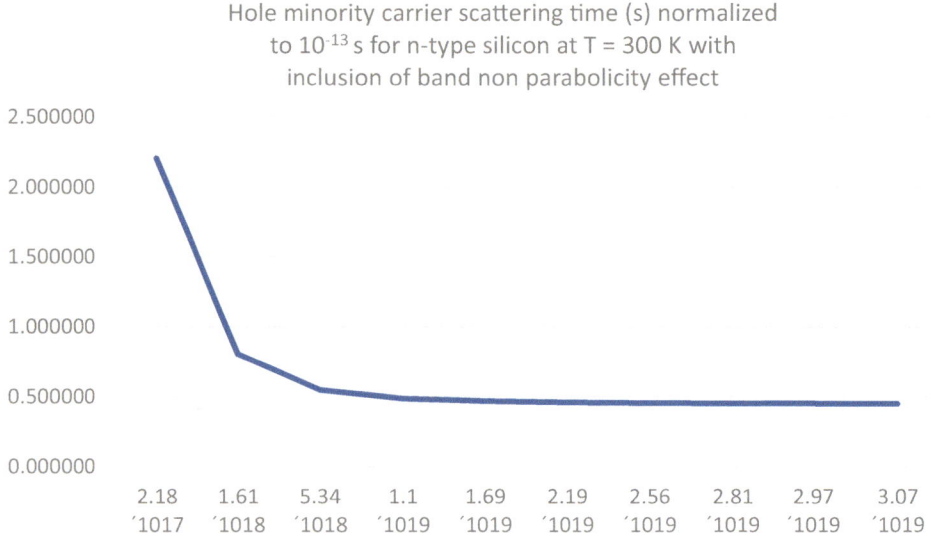

Fig. 3.9 Hole minority carrier scattering time normalized to 10^{-13} s in n-type phosphorous doped silicon at T = 300 K with inclusion of band non parabolicity

roughness scattering also. Whereas for minority carrier electron mobility, the substrate is p-type or boron doped, the interface is less smooth and has more dangling bonds than when the substrate is n-type or phosphorous doped which results in more interface passivation during oxide growth and more smoother interface and less number of dangling bonds. Obviously, this shows that due to reduced interface roughness scattering, minority carrier hole scattering time is larger than minority carrier electron scattering time. Besides, for p-FET devices on n-type silicon substrate, when the inversion hole channel is formed, the gate potential is negative and the vertical electric field emanates from the positively charged hole inversion layer and terminates at the negatively charged carriers on gate electrode due to negative potential. This results in redistribution of gate potential with the negatively charged electrons almost pinned to gate electrode and making the inversion channel thickness more uniform in p-FET with hole as minority carriers than in n-FET with electron as minority carriers. The uniform channel thickness even if reduced near drain side due to drain potential, also contributes to less surface roughness scattering as well as enhanced screening of inversion layer hole of ionized dopants in the depletion region when low field ionized impurity scattering is dominant.

Table 3.10 The relative increment ratio $\tau_{cp}{}^{minor}/\tau_{cp}{}^{maj}$ for hole transport

N_A or N_D (/cm^3)	$\tau_p{}^{minor}/\tau_p{}^{maj}$
2.18×10^{17}	1.8924
1.61×10^{18}	2.0458
5.34×10^{18}	2.0953
1.1×10^{19}	2.1136
1.69×10^{19}	2.1205
2.19×10^{19}	2.1237
2.56×10^{19}	2.125257
2.81×10^{19}	2.12614
2.97×10^{19}	2.12663
3.07×10^{19}	2.12686

Table 3.10 shows the relative increment ratio $\tau_{cp}{}^{minor}/\tau_{cp}{}^{maj}$ for hole transport taking either n-type or p-type silicon doping concentrations, from where we witness that the ratio sharply increases for lower doping concentrations or non degenerate doping concentrations-that are employed in today's FET devices but the ratio saturates at a nearly constant value for high degenerate doping for n-type and p-type silicon at T = 300 K. This saturating and the incremental values of the ratio are also larger in magnitude than Table 3.5 values for electron as carrier.

Figure 3.10 shows the plot of $\tau_p{}^{minor}/\tau_p{}^{maj}$ taking n-type or p-type substrate concentrations up to degenerate regime for silicon at T = 300 K. Higher increment and saturation value is noted in Fig. 3.10 when hole is both minority and majority carrier compared to Fig. 3.5 when electron is both minority and majority carrier in a FET.

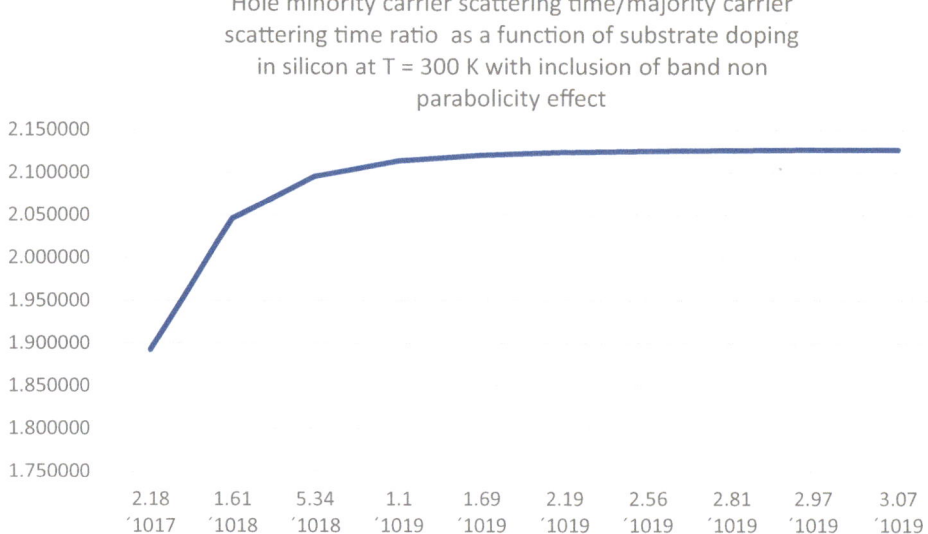

Hole minority carrier scattering time/majority carrier scattering time ratio as a function of substrate doping in silicon at T = 300 K with inclusion of band non parabolicity effect

Fig. 3.10 Hole minority carrier scattering time to majority scattering time ratio with respect to n-type or p-type substrate doping concentrations up to degenerate regime with inclusion of band non parabolicity. Even though majority transport based p-FETs and minority transport based p-FETs show lower inversion channel mobility than their counterpart for electron transports in n-FET, the scattering time for hole as majority and minority carriers enhancements suggest there is significant momentum or velocity gain for holes in transport before scattering events occur and if this momentum gain cannot subsequently get reduced by faster momentum relaxation time within successive number of scattering collisions, this may be lead to carrier temperature rise when the carriers are under high lateral field due to drain voltage, which suggests that p-FETs are more prone to self heating effects (SHE) than n-FETs

References

1. Low Temperature Electronics, Physics, Devices, Circuits and Applications, Edmundo A. Gutiérrez-D., M. Jamal Deen and C. Claeys, Academic Press, 2001.
2. Semiconductor Physics and Devices Basic Principles, Donald A. Neamen, Fourth Edition, McGraw Hill Companies Inc, 2012.
3. An Analytical, Temperature-dependent Model for Majority- and Minority-carrier Mobility in Silicon Devices, SUSANNA REGGIANI, MARINA VALDINOCI, LUIGI COLALONGO, MASSIMO RUDAN and GIORGIO BACCARANI, VLSI DESIGN, January 2000, Semiconductor Device Modeling and Simulation (section), Vol. 10, No. 4, pp. 467–483.
4. MOSFET Performance Scaling-Part I: Historical Trends, Ali Khakifirooz and Dimitri A. Antoniadis, IEEE Transactions on Electron Devices, Vol. 55, No. 6, June 2008, pp. 1391–1400.

Band Non Parabolicity Introduced T = 300 K Other Critical Parameters Derivations that Impact FET Performance

Electron majority carrier transport in majority carrier FET devices such as junctionless n-FET and n-high electron mobility transistors (HEMT) has some important device physical effect based illustrations. First in donor doped n-type material where the donors are positively ionized, the electrons as majority carriers experience an attractive Coulomb force field and increased directed collisions that significantly impacts and orient their momentum, drift velocity and mobility with further collisions. Also since there is no depletion region under the inversion channel like in minority carrier FET devices, these channel electrons are highly mobile, accumulates at the dielectric-semiconductor interface with dopants distribution surrounding them, as a result, inversion channel carrier screening of ionized dopants is non-existent in majority carrier transport based FET devices impacting mobility where the natural screening induced mobility enhancement by gate voltage overdrive found in inversion minority carrier dominant FETs, also does not happen in majority carrier FET devices either electron or hole based transport. In accumulation, due to inversion channel electrons in majority carrier FETs, although has higher surface roughness scattering but their inversion layer thickness does not get reduced to the extent we observe in minority carrier transport related FETs near drain contact when the drain bias voltage is high to introduce higher drift field for saturation region On-current which is what the benchmark parameter for semiconductor logic devices-a building block of computer microprocessors. Therefore as this layer in majority carrier FET device is not as squeezed as minority carrier FET devices under inversion bias and saturation regime operation, at high degenerate doping with near 100% ionization, carrier-carrier scattering, a feature when the device is in non equilibrium with thermal lattice temperature of the substrate, is reduced more in majority carrier FET than in minority carrier FET. We now systematically calculate first the majority carrier electron mobility in phosphorous doped

N. S. Ashraf, *Parameter-Centric Scaled FET Devices*, Synthesis Lectures on Emerging Engineering Technologies, https://doi.org/10.1007/978-3-031-84286-3_4

Table 4.1 Majority conductivity electron effective mass normalized to m_o, electron scattering time τ_n^{maj} (s) and mobility μ_n^{maj} (cm^2/V-s) with band band non parabolicity effect

N_D (/cm^3)	m_{cn}^{majr}/m_o	τ_n^{maj} (s)	μ_n^{maj} (cm^2/V-s)
2.18×10^{17}	0.29846	1.9942×10^{-13}	1173.503
1.61×10^{18}	0.3007	5.2403×10^{-14}	306.0725
5.34×10^{18}	0.30649	2.5949×10^{-14}	148.698
1.1×10^{19}	0.314515	1.9798×10^{-14}	110.556
1.69×10^{19}	0.3219	1.7749×10^{-14}	96.84
2.19×10^{19}	0.3274	1.6874×10^{-14}	90.519
2.56×10^{19}	0.33104	1.6445×10^{-14}	87.248
2.81×10^{19}	0.33327	1.6214×10^{-14}	85.473
2.97×10^{19}	0.3346	1.6094×10^{-14}	84.477
3.07×10^{19}	0.3354	1.6023×10^{-14}	83.904

n-type silicon when the substrate degeneracy effect on band non parabolicity is considered. We will need the Table 3.1 m_{cn}^{maj}/m_o values multiplied by free electron mass m_o and Table 3.2 τ_n^{maj} (s) and Eq. (3.3) from Chap. 3 of this book to arrive at majority carrier electron mobility μ_n (cm^2/V-s) as a function of the reference substrate doping (donor doped) at T = 300 K.

Table 4.1 lists these values.

Figure 4.1 shows the majority carrier mobility for electron in n-type phosphorous doped silicon with the inclusion of band non parabolicity as a function of substrate doping. The Drude mobility has advantage over Matthiessen's rule based part-by part scattering events based overall mobility computation in that the scattering time in Drude mobility is always overall value and the modeling approach to define analytical equation to derive these scattering times as discussed in Chap. 3 of this book, shows that taking some mobility point values experimental majority carrier electron mobility-doping concentration such as in [1], enables a design engineer to simply use the Drude mobility equation to calculate the mobility once the near accurate degeneracy effect induced non parabolicity induced electron majority conductivity electron mass as a function of doping can be derived which is also shown in Chap. 3 of this book.

For minority electron mobility in p-type boron doped silicon at T = 300 K, initially for low non degenerate doping regime, due to threshold voltage being low, enough inversion carriers are generated in the channel and the channel layer is also thicker with slightly increased gate field with low drain bias, making the case of ionized impurity dopants scattering screening possible resulting in inversion carrier mobility increase. But, as the doping goes from non degenerate to degenerate to high degenerate, vertical field or gate voltage requirements also shift to higher voltage to maintain inversion channel density due to increase of threshold voltage with substrate doping. Now, the ionized impurity scattering region for carrier transport is very short and the mobility is dominated by fast

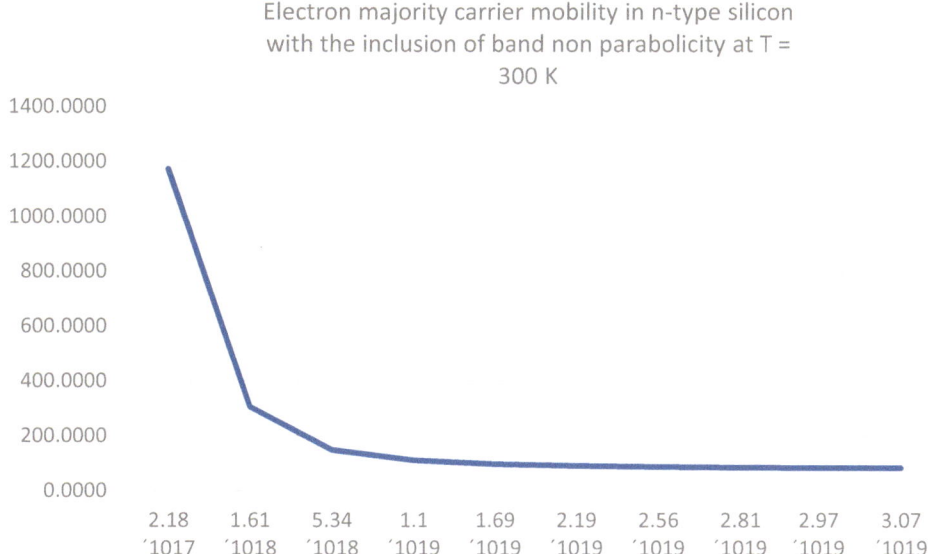

Fig. 4.1 Electron majority carrier mobility μ_n^{maj} (cm^2/V-s) in n-type phosphorous doped silicon at T = 300 K in degenerately doped substrate with inclusion of band non parabolicity effect and as we can see majority electron carrier mobility reaches a fluctuating significantly reduced constant value at high or extreme degenerate doping concentrations, as carrier to dopant scattering is more intense here due to attractive force field and also accumulated majority electron channel faces higher surface roughness scattering as doping is increased even if the drain bias is small or under low lateral field but under high vertical gate bias with significant accumulation of majority carrier channel layer surrounded by their native positively ionized donor dopants

reaching mobility peak by phonon limited mobility scattering at T = 300 K and subsequently, by optical phonon scattering and interface roughness scattering making mobility drastically reduced. Yet when compared with majority carrier electron mobility, the minority carrier electron mobility mostly benefits from screening of ionized impurity scattering and for both transport cases, the channel layer thickness for the same gate voltage and low drain voltage, has a different contribution through ionized impurity scattering events. Majority carrier also loses momentum frequently by direct attractive force-field induced collisions between negatively charged electrons and positively charged donors, but minority carriers in the vicinity of negatively charged acceptors in p-type silicon, experience a repulsive field, so direct momentum loss event does not happen in this case. Accumulation of majority carriers when they form inversion channel in majority carrier FET, also has higher surface roughness scattering than minority carrier based inversion channel case. Table 4.2 lists the value of minority carrier electron conduction mass normalized to m_0, m_{cn}^{minor} from Table 3.4 of Chap. 3 of this book and the minority carrier scattering time

Table 4.2 Listing of the value of minority carrier electron conduction mass normalized to m_O, the minority carrier scattering time τ_n^{minor} (s) and the overall Drude mobility equation based inversion channel mobility as computed

N_A (/cm^3)	m_{cn}^{minor}/m_O	τ_n^{minor} (s)	μ_n^{minor} (cm^2/V-s)
2.18×10^{17}	0.2984	1.965×10^{-13}	1156.552
1.61×10^{18}	0.30114	6.347×10^{-14}	370.1704
5.34×10^{18}	0.30809	4.076×10^{-14}	232.358
1.1×10^{19}	0.31772	3.5497×10^{-14}	196.222
1.69×10^{19}	0.3286	3.3747×10^{-14}	180.372
2.19×10^{19}	0.33318	3.2999×10^{-14}	173.945
2.56×10^{19}	0.33751	3.2633×10^{-14}	169.813
2.81×10^{19}	0.340164	3.244×10^{-14}	167.492
2.97×10^{19}	0.34175	3.2334×10^{-14}	166.16996
3.07×10^{19}	0.3427	3.2273×10^{-14}	165.3967

τ_n^{minor} (s) and the overall Drude mobility equation based inversion channel mobility as computed.

Figure 4.2 shows the value of inversion channel mobility for electron as minority carriers in p-type silicon at T = 300 K taking the effect of band non parabolicity.

Table 4.3 lists the increment ratio of μ_n (minority electron)/μ_n (majority electron) with the substrate doping N_D or N_A but the same subset as we have illustrated in this book with all figures and Tables. The benefit of speed of minority electron carrier based FET over majority electron carrier based FET is clearly visible from the Table 4.3.

Figure 4.3 shows the increment ratio $\mu_n^{minor}/\mu_n^{maj}$ for electron when transport in inversion channel n-FET and majority electron channel based junctionless FET and n-HEMT are considered. The benefit of minority electron transport based n-FET is clearly visible as with degeneracy, when the non parabolicity of conduction band is taken into account, the ratio progressively reaches a saturating value near 2 for the substrate doping being considered. The ratio itself does not show whole picture of detrimental outlook for FET operating with degenerate substrate doping when both minority and majority electron mobilities are reduced and interface roughness scattering almost drives these mobilities to their lowest ebb.

Even though actual mobilities for both inversion carrier and majority carrier electron are reduced at high degenerate substrate value, the inclusion of band non parabolicity with doping dependent increase of both minority and majority conductivity electron mass m_{cn}, make these mobilities to remain almost constant at these degenerate substrate doping and the rate of decrease of mobility for both majority and minority case are slower, hence based on the residual saturating decreased value plateaus of these mobilities, the ratio shows an increasing trend that we can gain some transport advantage by operating FET

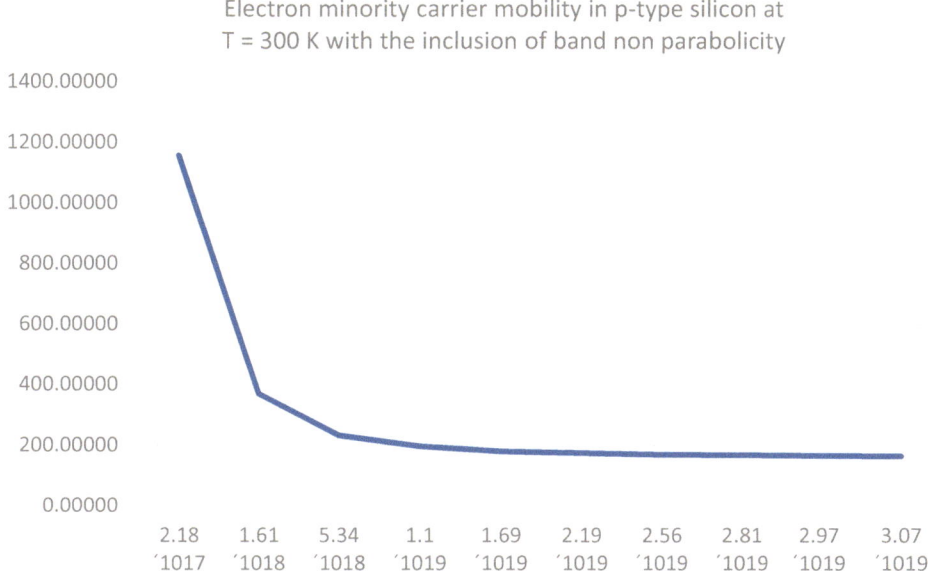

Fig. 4.2 Electron minority carrier mobility (cm^2/V-s) in p-type boron doped silicon at T = 300 K taking the substrate doping into the degenerate regime with inclusion of band non parabolicity effect

Table 4.3 Increment ratio of μ_n (minority electron)/μ_n (majority electron)

N$_A$ or N$_D$ (/cm^3)	μ_n^{maj} (cm^2/V-s)	μ_n^{minor} (cm^2/V-s)	μ_n^{minor}/ μ_n^{maj}
2.18×10^{17}	1173.503	1156.552	0.98555
1.61×10^{18}	306.0725	370.1704	1.2094
5.34×10^{18}	148.698	232.358	1.5626
1.1×10^{19}	110.556	196.222	1.77487
1.69×10^{19}	96.84	180.372	1.86258
2.19×10^{19}	90.519	173.945	1.9216
2.56×10^{19}	87.248	169.813	1.9463
2.81×10^{19}	85.473	167.492	1.9596
2.97×10^{19}	84.477	166.16996	1.96704
3.07×10^{19}	83.904	165.3967	1.97126

devices near degenerate substrate doping when building n-FET over junctionless FET in logic devices, but only in relative comparability viewpoint, but not with respect to the actual mobility advantage scenario which aggravates with increased substrate doping up to degenerate doping for both majority and minority carrier electrons.

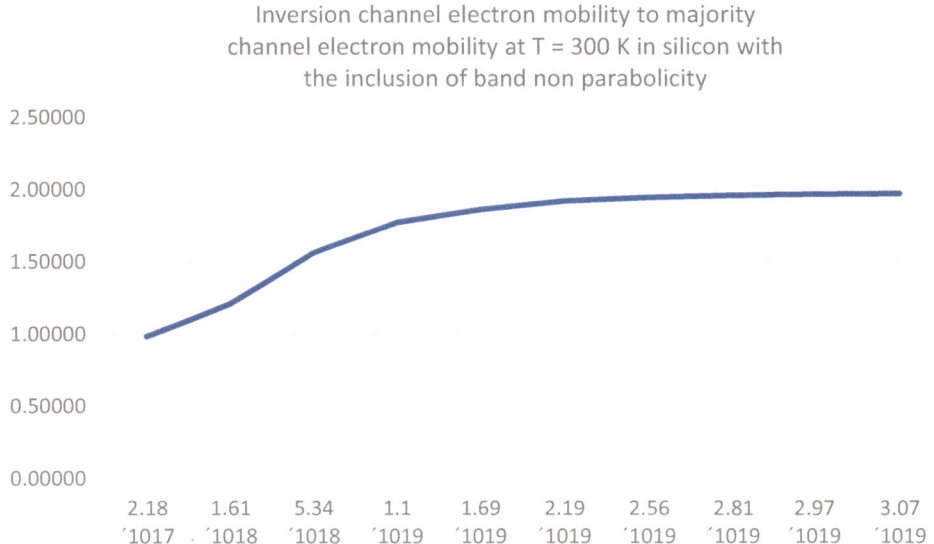

Fig. 4.3 The mobility increment ratio $\mu_n^{minor}/\mu_n^{maj}$ for electron when transport in inversion channel n-FET and majority electron channel based junctionless FET and n-HEMT are considered

For hole majority carrier mobility in p-type silicon, since the majority carrier conductivity hole effective mass quite a proportionate larger than electron majority carrier conductivity effective mass, the majority carrier hole mobility in boron doped p-type silicon substrate at T = 300 K gets increasingly attenuated as doping of the substrate is increased to degenerate regime. But near high substrate doping for both n-type and p-type silicon, majority electron mobility and majority hole mobility values are comparable factor of each other even though majority hole mobility goes down in values. As a result, the scattering time for majority hole carrier increases in proportion to majority electron carrier transport scattering time in n-silicon, although effects of surface roughness scattering are more intense and equally likely, the only difference is the higher ionization percentage of p-type silicon over n-type silicon. We now use the values of m_{cp}^{maj} normalized to m_o and scattering time for majority carrier hole τ_p^{maj} (s) of the Chap. 3 of this book of Table 3.7 to list the values of majority carrier hole mobility in p-type silicon at T = 300 K in the following Table 4.4.

Figure 4.4 shows the majority carrier hole mobility in p-type boron doped silicon substrate for T = 300 K with the inclusion of band non parabolicity effect. Hole mobilities are largely attenuated for majority carrier hole transport in p-type substrate. The force field action of attractive force between hole in transport and an acceptor dopant in the vicinity of the depletion region in p-type silicon is different than the force field action of the attractive force of electron in transport with a lighter conductivity mass and a donor dopant in the vicinity of the depletion region in n-type silicon, so after collision momentum

Table 4.4 Listing of the values of m_{cp}^{maj} normalized to m_o, scattering time for majority carrier hole τ_p^{maj} (s) and majority carrier hole mobility in p-type silicon

N_A (/cm^3)	m_{cp}^{maj}/m_o	τ_p^{maj} (s)	μ_p^{maj} (cm^2/V-s)
2.18×10^{17}	0.54451	1.1656×10^{-13}	375.963
1.61×10^{18}	0.55027	3.932×10^{-14}	125.499
5.34×10^{18}	0.56498	2.613×10^{-14}	81.228
1.1×10^{19}	0.58537	2.3074×10^{-14}	69.23
1.69×10^{19}	0.60413	2.2058×10^{-14}	64.126
2.19×10^{19}	0.61803	2.1624×10^{-14}	61.4509
2.56×10^{19}	0.62713	2.1412×10^{-14}	59.965
2.81×10^{19}	0.63272	2.12999×10^{-14}	59.125
2.97×10^{19}	0.63606	2.1238×10^{-14}	58.643
3.07×10^{19}	0.63804	2.1203×10^{-14}	58.365

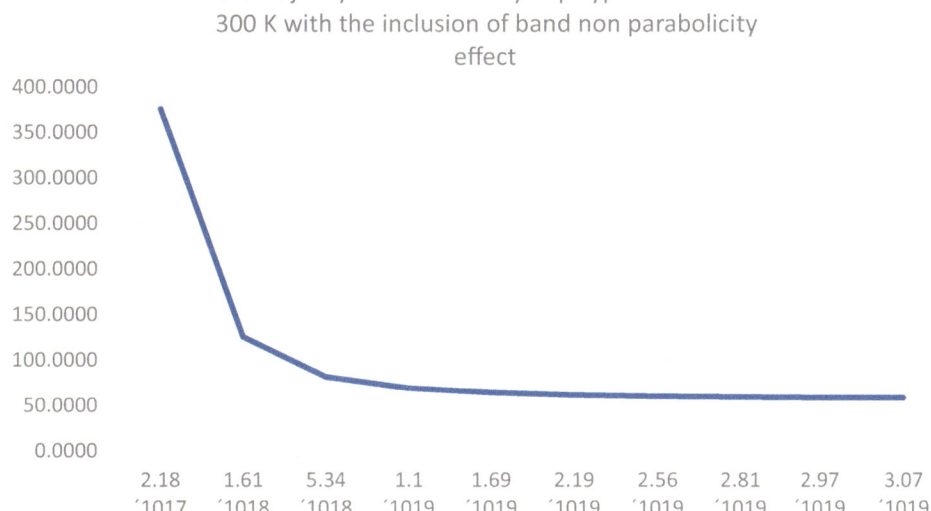

Fig. 4.4 Hole majority carrier mobility in cm^2/V-s in p-type boron doped silicon substrate at T = 300 K with the inclusion of band non parabolicity effect

distribution effects will be different for majority hole transport in p-type silicon compared to majority electron transport in n-type silicon at T = 300 K.

For minority carrier hole mobility, analogous to minority carrier electron mobility in p-type silicon substrate at T = 300 K with the inclusion of band non parabolicity, minority carrier holes in n-type donor doped silicon substrate at T = 300 K, also form inversion layer induced screening of ionized donor dopants in the depletion region, but due to

Table 4.5 Listing of minority carrier hole conductivity mass m_{cp}^{minor} normalized to m_o, minority carrier hole scattering time τ_p^{minor} (s) and minority carrier hole mobility at T = 300 K

N_D (/cm^3)	m_{cp}^{minor}/m_o	τ_p^{minor} (s)	μ_p^{minor} (cm^2/V-s)
2.18×10^{17}	0.5444	2.20578×10^{-13}	711.616
1.61×10^{18}	0.5494	8.0439×10^{-14}	257.146
5.34×10^{18}	0.56228	5.47498×10^{-14}	171.0139
1.1×10^{19}	0.58010	4.8769×10^{-14}	147.653
1.69×10^{19}	0.59651	4.6774×10^{-14}	137.717
2.19×10^{19}	0.6087	4.5923×10^{-14}	132.504
2.56×10^{19}	0.61869	4.5506×10^{-14}	129.1806
2.81×10^{19}	0.6216	4.52865×10^{-14}	127.9556
2.97×10^{19}	0.6245	4.51653×10^{-14}	127.0206
3.07×10^{19}	0.62629	4.50959×10^{-14}	126.463

difference in non parabolicity factor due to degeneracy, holes as minority carriers have different conductivity mass in donor doped substrate than what we have seen Table 4.4 for majority carrier holes. The inversion layer thickness for minority carrier holes in n-type substrate will be slightly larger as phosphorous dopants are slightly less ionized than boron acceptor dopants in silicon at T = 300 K. As a result, for this reason, a little bit larger screening for hole minority carriers will result in hole minority carrier mobility to be quite a factor larger than hole majority carrier mobility but these mobilities are still smaller in magnitude than minority and majority carrier electron mobility as hole conductivity effective mass at T = 300 K for both type of transport are quite a factor larger than electron conductivity effective mass at T = 300 K. From Table 3.9 of Chap. 3 of this book, the minority carrier hole conductivity mass m_{cp}^{minor} normalized to mo, the minority carrier hole scattering time τ_p^{minor} (s) are listed along with computed minority carrier hole mobility at T = 300 K in phosphorous doped n-type silicon from Drude mobility Eq. (3.3 with the m_{cp}^{minor} and τ_p^{minor} and this data is shown in Table 4.5.

We have note when the substrate doping becomes highly degenerate, both minority carrier electron mobility and majority carrier electron mobility values get closer to each other and as well substantially reduced in the vicinity of some 100 cm^2/V-s. So with high lateral field and simultaneously vertical gate field required for degenerately doped minority carrier n-FET and p-FET, the speed of the devices and therefore on current magnitudes become closer to each other due to similarities in mobility values at these doping levels with no advantage accrued in terms of saturated drift velocity and maximum magnitude of On current gain perspectives at T = 300 K. Figure 4.5 shows the minority carrier hole mobility as a function of donor doped substrate concentrations in n-type silicon with the inclusion of band non parabolicity effect at T = 300 K.

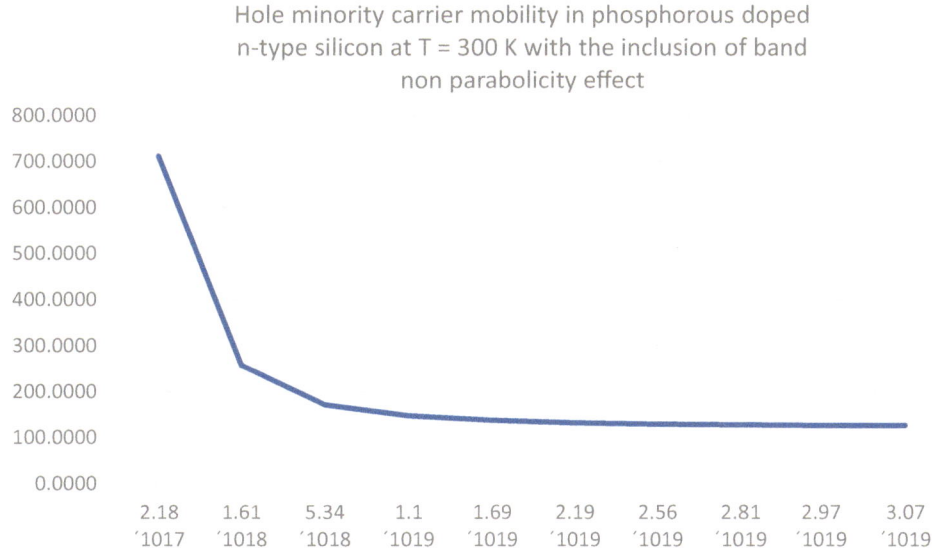

Fig. 4.5 Hole minority carrier mobility μ_p^{minor} (cm^2/V-s) in phosphorous doped n-type silicon substrate at T = 300 K with the inclusion of band non parabolicity effect

Table 4.6 shows the increment ratio for $\mu_p^{minor}/\mu_p^{maj}$ for either p-type or n-type substrate where the hole carriers are majority or minority. We again witness progressively higher ratio magnitude with the substrate doping values with the similar saturating tendency of the ratio values near very high degenerate doping in either substrate. So, even though the mobility values are quite lower in respective majority carrier and minority carrier FET devices, this Table 4.6 and associated Fig. 4.6 show that if degenerate doping in the substrate is mandatory, sometimes from the observation of prevention of source and drain depletion region merging in the substrate when the channel length is aggressively scaled, minority carrier hole p-FET will get a speed advantage and on current magnitude advantage over majority carrier hole p-FET and junctionless p-FET.

Figure 4.6 shows the relative increment ratio $\mu_p^{minor}/\mu_p^{maj}$ for either donor or acceptor doped silicon substrate where the carriers are either majority or minority, with substrate doping concentration variation up to degenerate level including the effect of band non parabolicity.

Effective intrinsic carrier concentration in n-type and p-type silicon at T = 300 K under band non parabolicity and ionized dopants induced band gap narrowing

The need to compute the effective intrinsic carrier concentration in n-type and p-type silicon at T = 300 K up to degenerate substrate doping concentrations under non parabolicity and non parabolicity coupled with ionized dopants induced band gap narrowing at T = 300 K in n-type and p-type silicon substrate, was the focus of this chapter as previous

Table 4.6 Shows the increment ratio for $\mu_p^{minor}/\mu_p^{maj}$ for either p-type or n-type substrate where the hole carriers are majority or minority

N_A or N_D (/cm^3)	μ_p^{maj} (cm^2/V-s)	μ_p^{minor} (cm^2/V-s)	$\mu_p^{minor}/\mu_p^{maj}$
2.18×10^{17}	375.963	711.616	1.89278
1.61×10^{18}	125.499	257.146	2.04899
5.34×10^{18}	81.228	171.0139	2.105356
1.1×10^{19}	69.23	147.653	2.13279
1.69×10^{19}	64.126	137.717	2.14760
2.19×10^{19}	61.4509	132.504	2.15626
2.56×10^{19}	59.965	129.1806	2.15427
2.81×10^{19}	59.125	127.9556	2.16415
2.97×10^{19}	58.643	127.0206	2.165998
3.07×10^{19}	58.365	126.463	2.16676

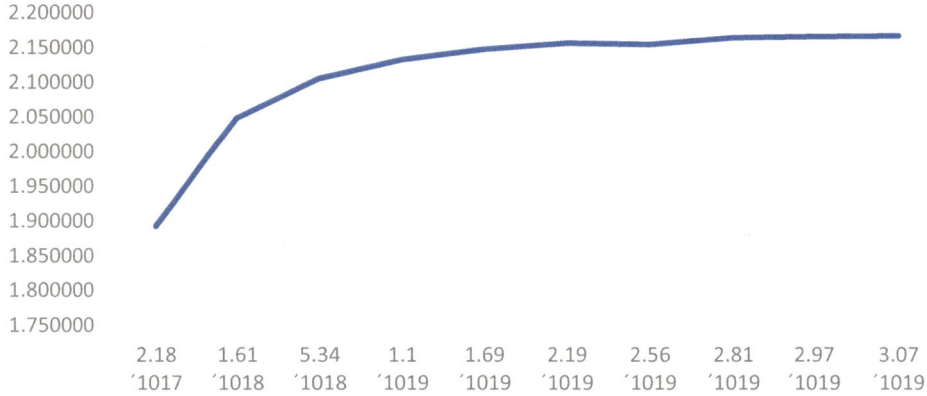

Fig. 4.6 Hole minority carrier mobility to majority carrier mobility ratio as a function of substrate doping concentrations up to degenerate level including band non parabolicity. The difference in Fig. 4.6 from Fig. 4.3 where electron minority to majority mobility ratio was plotted is that the ratio values for Table 4.6 are concentrated around small increments in 2 which makes the curve for Fig. 4.6 showing more sharper trend than Fig. 4.3 and therefore, even though the mobility values for hole based FET are quite smaller than electron based FET excepting the degenerate doping regime, the higher ratio of Table 4.6 and Fig. 4.6 show that hole carrier based minority p-FET over majority p-FET will offer better advantage for speed and drive current enhancement compared to electron carrier based minority n-FET over majority n-FET

referenced did not account for these physical effects accurately [2–4]. We now first compute the effective intrinsic carrier concentration in n-type silicon with phosphorous doping up to degenerate substrate doping at T = 300 K under band non parabolicity and band non parabolicity coupled with ionized donor dopants induced band gap narrowing from equations in [5]. Band gap narrowing decreases the minima of conduction band E_c (eV) at T = 300 K in n-type silicon by ΔE_c (eV) [5] and increases reference $E_V = 0$ eV at T = 300 K to ΔE_v (eV) [5]. We will need the $N_D{}^+$ and η_c values from Table 4.15 of Chap. 2 and η_v which has been previously referred in Chaps. 1 and 2, needs to be recomputed by incorporation of ionized dopant induced band gap narrowing of ΔE_c and ΔE_v. The free carrier ionized dopant density in n-type silicon is $n_o = N_D{}^+$, where band non parabolicity induced ionized dopants are given Table 4.15 of Chap. 2. The rest of the equations to calculate effective intrinsic carrier concentration $n_i{}^{eff}$ (/cm^3) in n-type silicon at T = 300 K with doping dependent band gap narrowing and under band non parabolicity effect are:

$$\Delta E_c(N_D) = -14.84 \times 10^{-3} \left(\frac{N_D^+}{10^{18}} \right)^{\frac{1}{3}} + 0.78 \times 10^{-3} \left(\frac{N_D^+}{10^{18}} \right)^{1/2} \tag{4.1}$$

$$\Delta E_v(N_D) = 15.08 \times 10^{-3} \left(\frac{N_D^+}{10^{18}} \right)^{\frac{1}{4}} + 0.74 \times 10^{-3} \left(\frac{N_D^+}{10^{18}} \right)^{1/2} \tag{4.2}$$

$$\eta_v = -\left[\frac{1.12 - [|\Delta E_c(N_D)| + \Delta E_v(N_D)]}{kT} \right] - \eta_c \tag{4.3}$$

$$p_o = N_v e^{\eta_v} \tag{4.4}$$

In the calculation of N_V, since n-type donor degeneracy does not affect the minority carrier concentration which is extremely low value and non degenerate order, we can use N_v quoted in [6] and [7] to be 1.83×10^{19}/cm^3 with $m_p = 0.81\ m_o$ which has been shown in Chap. 1 discussion. Finally,

$$n_i^{eff} = \sqrt{n_o p_o} \tag{4.5}$$

Table 4.7 lists the related parameters with the final calculation of $n_i{}^{eff}$ (/cm^3) with its normalized value to 10^{10}/cm^3.

Figure 4.7 shows the plot of $n_i{}^{eff}/10^{10}$ (/cm^3) with the incorporation of non parabolicity induced alterations in $N_D{}^+$ and ionized dopant induced band gap narrowing for n-type phosphorous doped silicon at T = 300 K as relative to bulk original reference substrate doping density of n-type silicon.

For calculation of effective intrinsic carrier concentration $n_i{}^{eff}$ (/cm^3) in boron doped p-type silicon at T = 300 K for considering doping up to degenerate values, the preceding analysis of effective intrinsic carrier concentration computation for phosphorous doped n-type silicon at T = 300 K will apply and the band non parabolicity in the degenerately

Table 4.7 The values of n_i^{eff} (/cm^3) with its normalized value to 10^{10}/cm^3 at T = 300 K for n-type silicon with band non parabolicity

N_D (/cm^3)	N_D^+ (/cm^3)	η_c	η_v	n_i^{eff} (/cm^3)	$n_i^{eff}/10^{10}$(/cm^3)
2.18×10^{17}	2.033×10^{17}	-5.0656	-37.4915	1.3935×10^{10}	1.3935
1.61×10^{18}	1.3035×10^{18}	-3.212	-38.829	1.808×10^{10}	1.808
5.34×10^{18}	4.5635×10^{18}	-1.969	-39.5355	2.3765×10^{10}	2.3765
1.1×10^{19}	1.0177×10^{19}	-1.171	-39.835	3.0547×10^{10}	3.0547
1.69×10^{19}	1.623×10^{19}	-0.7034	-39.966	3.613×10^{10}	3.613
2.19×10^{19}	2.1345×10^{19}	-0.4274	-40.022	4.0291×10^{10}	4.0291
2.56×10^{19}	2.511×10^{19}	-0.2631	-40.0453	4.319×10^{10}	4.319
2.81×10^{19}	2.765×10^{19}	-0.1654	-40.0564	4.507×10^{10}	4.507
2.97×10^{19}	2.9269×10^{19}	-0.1081	-40.063	$4.6223 \cdot \times 10^{10}$	4.6223
3.07×10^{19}	3.0283×10^{19}	-0.07226	-40.0665	4.6935×10^{10}	4.6935

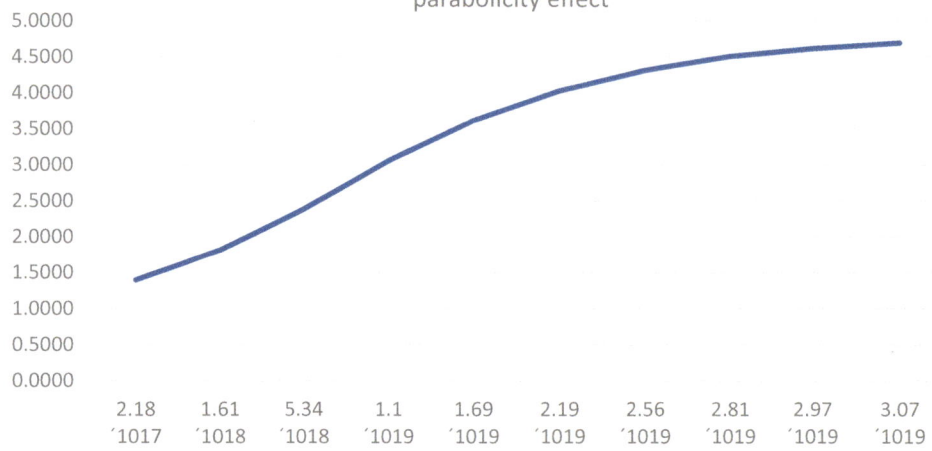

Effective intrinsic carrier concentration (/cm^3) normalized to 10^{10}(/cm^3) for n-type phosphorous doped silicon at T = 300 K including band non parabolicity effect

Fig. 4.7 The effective intrinsic carrier concentration n_i^{eff}/cm^3 normalized to 10^{10}/cm^3 for n-type phosphorous doped silicon at T = 300 K under band non parabolicity and band non parabolicity altered actual ionized dopant induced band gap narrowing. The original set of bulk substrate doping data are taken as reference for X-axis

doped p-type silicon will alter the actual ionized acceptor doping density and also impact the ionized acceptor dopant induced band gap narrowing on first valence band maxima with ΔE_V (eV) upshift values (equation from [5]) and ΔE_c downshift values (equation from [5]), remembering that for p-type silicon, the conduction band profile in the vicinity of the minima will not be altered by band non parabolicity effect as it affects only the band structure that is majority carrier related. Free ionized carrier density in this case for p-type silicon is $p_o = N_A^-$ and we take the assistance of Table 4.18 of Chap. 2 of this book to collect the values of N_A^- and η_v along with the substrate reference doping set used throughout this book N_A (/cm^3). The new equations set are provided below:

$$\Delta E_v(N_A) = 18.46 \times 10^{-3} \left(\frac{N_A^-}{10^{18}}\right)^{\frac{1}{3}} - 2.63 \times 10^{-3} \left(\frac{N_A^-}{10^{18}}\right)^{1/2} \tag{4.6}$$

$$\Delta E_c(N_A) = -16.27 \times 10^{-3} \left(\frac{N_A^-}{10^{18}}\right)^{\frac{1}{4}} - 0.18 \times 10^{-3} \left(\frac{N_A^-}{10^{18}}\right)^{1/2} \tag{4.7}$$

$$\eta_c = -\left[\frac{1.12 - [\Delta E_v(N_A) + |\Delta E_c(N_A)|]}{kT}\right] - \eta_v \tag{4.8}$$

$$n_o = N_c e^{\eta_c} \tag{4.9}$$

In (4.9), the N_c value is 3.23×10^{19}/cm^3 with effective DOS electron mass $m_n = 1.18$ m_o at T = 300 K. Finally, Eq. (4.5) will be needed again to separately calculate n_i^{eff} for p-type boron doped silicon at T = 300 K up to degenerate doping concentrations and with band non parabolicity effect.

Table 4.8 shows this values systematically computed and arranged as relation to Table 4.7.

Figure 4.8 shows the trend of n_i^{eff} for boron doped p-type silicon at T = 300 K with consideration of degenerate doping values and band non parabolicity effect.

Now, we construct Table 4.9 where we assign the n_i^{eff} values of Table 4.7 as n_i^{effn} and Table 4.8 n_i^{eff} values as n_i^{effp} and list the ratio of n_i^{effn}/n_i^{effp} to show that how band non parabolicity marginally increases the effective carrier concentration of n-type silicon over effective carrier concentration of p-type silicon when proper degeneracy induced band non parabolicity effect is incorporated in the computation process.

Figure 4.9 shows the substrate type dependent effective intrinsic carrier concentration relative ratio of n_i^{effn}/n_i^{effp} taking the case of reference set of doping concentration data for n-type and p-type silicon at T = 300 K under band non parabolicity effect.

Table 4.8 The values of n_i^{eff} (/cm^3) with its normalized value to 10^{10}/cm^3 at T = 300 K for n-type silicon with band non parabolicity

N_A (/cm^3)	N_A^- (/cm^3)	η_v	η_c	n_i^{eff} (/cm^3)	$n_i^{eff}/10^{10}$(/cm^3)
2.18×10^{17}	1.8254×10^{17}	-4.6027	-37.9062	1.426×10^{10}	1.426
1.61×10^{18}	1.11716×10^{18}	-2.792	-39.206	1.8416×10^{10}	1.8416
5.34×10^{18}	4.23539×10^{18}	-1.4476	-39.9762	2.44×10^{10}	2.44
1.1×10^{19}	1.021155×10^{19}	-5239	-40.394	3.074×10^{10}	3.074
1.69×10^{19}	1.63859×10^{19}	-0.01108	-40.581	3.546×10^{10}	3.546
2.19×10^{19}	2.15169×10^{19}	0.2889	-40.675	3.877×10^{10}	3.877
2.56×10^{19}	2.52824×10^{19}	0.468455	-40.7256	4.0979×10^{10}	4.0979
2.81×10^{19}	2.78167×10^{19}	0.57522	-40.754	4.238×10^{10}	4.238
2.97×10^{19}	2.94354×10^{19}	0.6387	-40.770	4.3246×10^{10}	4.3246
3.07×10^{19}	3.0446×10^{19}	0.677	-40.78	4.3962×10^{10}	4.3962

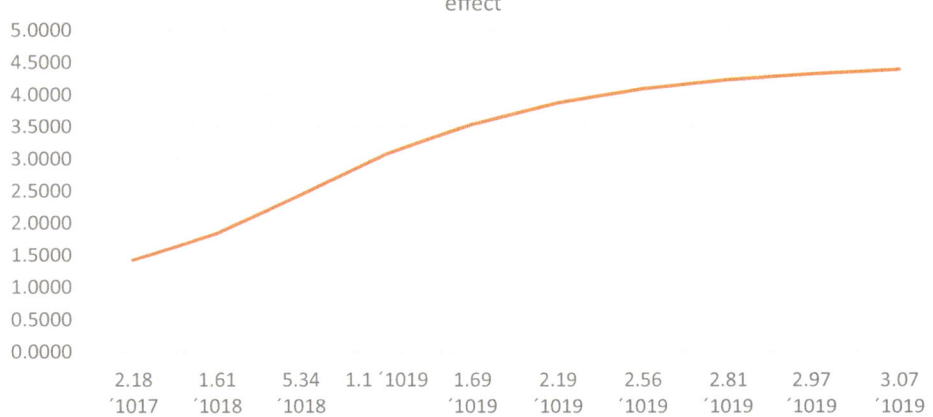

Effective intrinsic carrier concentration (/cm^3) normalized to 10^{10}(/cm^3) for p-type boron doped silicon at T = 300 K including band non parabolicity effect

Fig. 4.8 Effective intrinsic carrier concentration n_i^{eff} normalized to 10^{10}/cm^3 for p-type boron doped silicon considering the effect of band non parabolicity

Extrinsic Fermi energy E_F (eV) in n-type silicon and also in p-type silicon for consideration$ up to degenerate substrate doping and under band non parabolicity effect

From the ionized doping dependent under band non parabolicity effect, for n-type phosphorous doped silicon at T = 300 K, from listed values of η_c in Table 4.7, we can

Table 4.9 The ratio of n_i^{effn}/n_i^{effp} for silicon at T = 300 K with band non parabolicity

N_A (/cm³) (p-type), N_D (/cm³) (n-type)	n_i^{effn}/n_i^{effp}
2.18×10^{17}	0.9772
1.61×10^{18}	0.98175
5.34×10^{18}	0.97398
1.1×10^{19}	0.99372
1.69×10^{19}	1.01889
2.19×10^{19}	1.03923
2.56×10^{19}	1.05395
2.81×10^{19}	1.06347
2.97×10^{19}	1.06884
3.07×10^{19}	1.067627

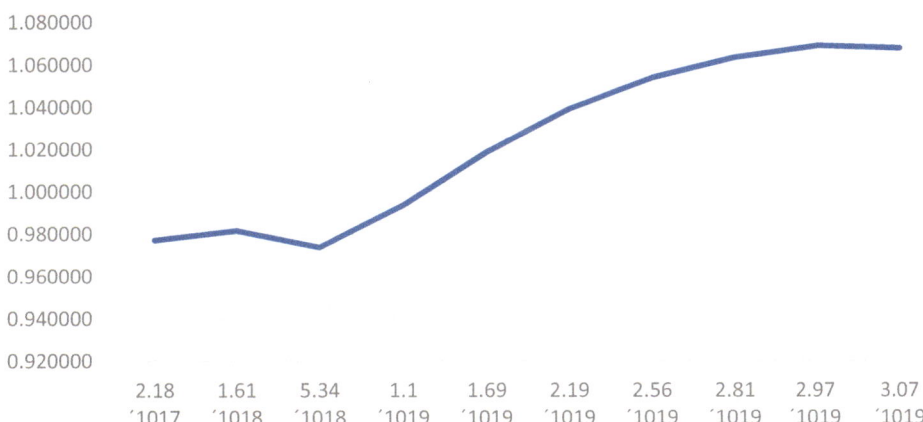

Ratio of effective intrinsic carrier concentration for n-type silicon to effective intrinsic carrier concentration for p-type silicon at T = 300 K with inclusion of band non parabolicity

Fig. 4.9 Ratio of effective intrinsic carrier concentration in n-type silicon to p-type silicon at T = 300 K when degenerate substrate doping region values are considered and band non parabolicity effects are incorporated

determine the extrinsic Fermi energy E_F (eV) for n-type silicon at T = 300 K under degenerate doping considerations and band non parabolicity effect from the following equation:

$$\eta_c = \frac{E_F - (1.12 - |\Delta E_c(N_D)|)}{kT} \tag{4.10}$$

Table 4.10 Fermi energy value E_F (eV) for n-type silicon with band non parabolicity

N_D (/cm^3)	η_c	$|\Delta E_c\ (N_D)|$ (eV)	E_F (eV)
2.18×10^{17}	-5.0656	8.3743×10^{-3}	0.98055
1.61×10^{18}	-3.212	0.0153	1.0216
5.34×10^{18}	-1.969	0.02295	1.0461
1.1×10^{19}	-1.171	0.02967	1.06
1.69×10^{19}	-0.7034	0.03443	1.06737
2.19×10^{19}	-0.4274	0.037556	1.0714
2.56×10^{19}	-0.2631	0.03955	1.0736
2.81×10^{19}	-0.1654	0.04077	1.07495
2.97×10^{19}	-0.1081	0.0415	1.0757
3.07×10^{19}	-0.07226	0.04196	1.0762

Here, E_c for non degenerate doping condition at T = 300 K for silicon is 1.12 eV [6–8] and $\Delta E_c\ (N_D)$ can be computed from Eq. (4.1). Table 4.10 lists the values of reference substrate doping for n-type phosphorous doped silicon at T = 300 K, η_c values from Table 4.7 and computed E_F (eV) from Eq. (4.10). Although from Eq. (4.1), $\Delta E_c\ (N_D)$ values are negative, signaling E_c minima is pushed down as doping becomes degenerate, for evaluation purpose as was done for $n_i{}^{eff}$ and η_v values Eqs. (4.10) and (4.3), $\Delta E_c\ (N_D)$ will be taken as absolute value to be deducted in Eq. (4.10) and also listed as absolute value in Table 4.10. Later we will follow this process for determining the E_F (eV) for degenerately boron doped p-type silicon at T = 300 K. but for this case in p-type silicon Eq. (4.6) reveals that ΔE_v is a growing positive value with p-type substrate doping up to degenerate doping values.

Figure 4.10 shows the trend of E_F (eV) in n-type phosphorous doped silicon at T = 300 K for doping up to degenerate values and under band non parabolicity.

For p-type boron doped silicon, the Fermi energy E_F (eV) derivation follows the same procedure described for the derivation steps of E_F in n-type phosphorous doped silicon at T = 300 K considering doping up to degenerate levels and under band non parabolicity. We will need the values of η_v from Table 4.8 and $\Delta E_V\ (N_A)$ values computed from Eq. (4.6) and now we list the equation from which values for E_F relevant for p-type silicon can be derived:

$$\eta_v = \frac{\Delta E_V (N_A) - E_F}{kT} \tag{4.11}$$

Table 4.11 lists the values of p-type silicon reference substrate doping up to degenerate values, η_v values and $\Delta E_V\ (N_A)$ values along with computed E_F in eV.

Figure 4.11 shows the variation of E_F (eV) in boron doped p-type silicon with substrate doping up to degenerate levels and with inclusion of band non parabolicity effect.

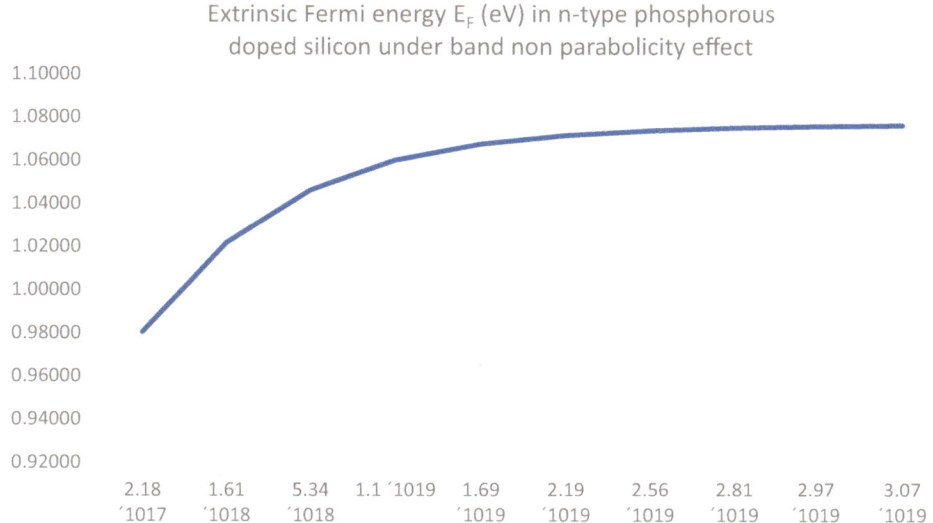

Extrinsic Fermi energy E_F (eV) in n-type phosphorous doped silicon under band non parabolicity effect

Fig. 4.10 Extrinsic Fermi Energy E_F (eV) in phosphorous doped n-type silicon for doping levels up to degenerate values and under band non parabolicity impacting both dopant ionization and dopant ionization induced band gap shift from peak conduction band minima value which is referenced at a non degenerate doping setting at T = 300 K

	N_A (/cm^3)	η_v	ΔE_v (N_A) (eV)	E_F (eV)
Table 4.11 Fermi energy value E_F (eV) for p-type silicon at T = 300 K	2.18×10^{17}	-4.6027	9.346×10^{-3}	0.12847
	1.61×10^{18}	-2.792	0.01637	0.0886
	5.34×10^{18}	-1.4476	0.02446	0.06192
	1.1×10^{19}	-5239	0.0316	0.04516
	1.69×10^{19}	-0.01108	0.03624	0.03653
	2.19×10^{19}	0.2889	0.03914	0.031665
	2.56×10^{19}	0.468455	0.040956	0.02883
	2.81×10^{19}	0.57522	0.04206	0.02718
	2.97×10^{19}	0.6387	0.04273	0.0262
	3.07×10^{19}	0.677	0.04313	0.0256

Band gap energy $E_G{}^n{}_{,BGN}$ (eV) for n-type phosphorous doped silicon at T = 300 K for doping levels up to degenerate concentrations and under band non parabolicity impacted dopant ionization and ionized dopant induced conduction band down shift

Now the procedure for calculation of band gap energy $E_G{}^n{}_{,BGN}$ (eV) for n-type phosphorous doped silicon at T = 300 K for doping levels up to degenerate doping concentrations

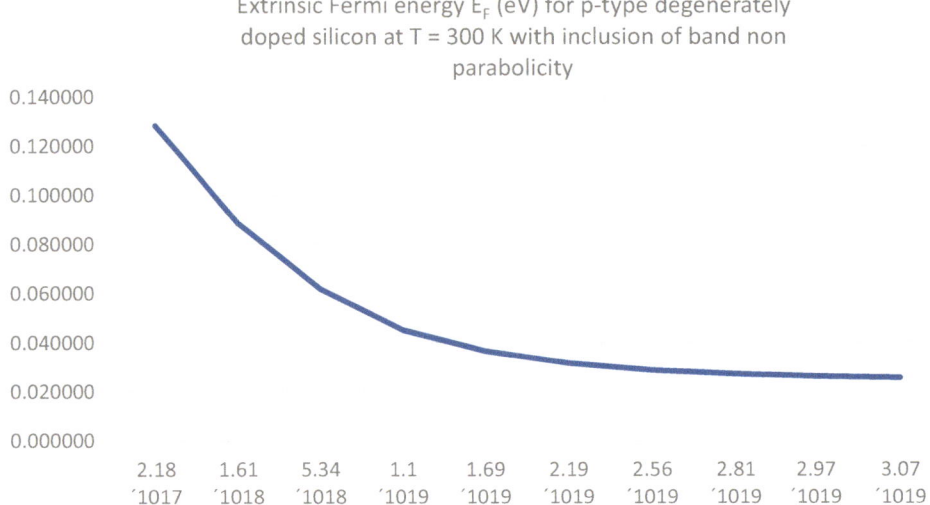

Fig. 4.11 Extrinsic Fermi energy level E_F (eV) for boron doped p-type silicon at T = 300 K with doping up to degenerate values considering band non parabolicity with its impact on acceptor ionization and acceptor ionization induced band gap narrowing effect on valence band maxima and conduction band minima

and under band non parabolicity impacted dopant ionization and ionized dopant induced conduction band down shift and valence band up shift. First will need calculation of ΔE_c (N_D) (eV) from Eq. (4.1) and ΔE_v (N_D) (eV). Now, parameter $E_G{}^n$,BGN (eV) is:

$$E^n_{G,BGN} = 1.12 - (|\Delta E_c(N_D)| + \Delta E_V(N_D)) \qquad (4.12)$$

Now Table 4.12 lists the related parameters along with substrate donor doping concentrations with the computed $E_G{}^n$,BGN (eV) values.

Figure 4.12 shows the value of $E_G{}^n$,BGN (eV) for phosphorous doped n-type silicon at T = 300 K with inclusion of band non parabolicity on dopant ionization and dopant ionization based band gap narrowing.

Now, we follow the same derivation method for band gap energy (eV) in boron doped p-type silicon $E_G{}^p$,BGN (eV) up to degenerate substrate doping concentrations with the inclusion of band gap narrowing and its impact on acceptor ionization and ionized acceptor induced valence band up shift and conduction band down shift: The relevant equation is given below:

$$E^p_{G,BGN} = 1.12 - (|\Delta E_c(N_A)| + \Delta E_V(N_A)) \qquad (4.13)$$

ΔE_V (N_A) (eV) in Eq. (4.13) can be calculated from Eq. (4.6) and ΔE_c (N_A) (eV) in Eq. (4.13) can be calculated from Eq. (4.7).

Table 4.12 Actual band gap in n-type silicon at T = 300 K E_G^n,BGN (eV) with band non parabolicity

N_D (/cm^3)	$|\Delta E_c (N_D)|$ (eV)	$\Delta E_V (N_D)$ (eV)	E_G^n,,BGN (eV)
2.18×10^{17}	8.3743×10^{-3}	0.01046	1.10117
1.61×10^{18}	0.0153	0.0169	1.0878
5.34×10^{18}	0.02295	0.02362	1.07343
1.1×10^{19}	0.02967	0.02929	1.06104
1.69×10^{19}	0.03443	0.03325	1.05232
2.19×10^{19}	0.037556	0.03582	1.046624
2.56×10^{19}	0.03955	0.03747	1.04298
2.81×10^{19}	0.04077	0.03849	1.04074
2.97×10^{19}	0.0415	0.03908	1.03942
3.07×10^{19}	0.04196	0.03945	1.03859

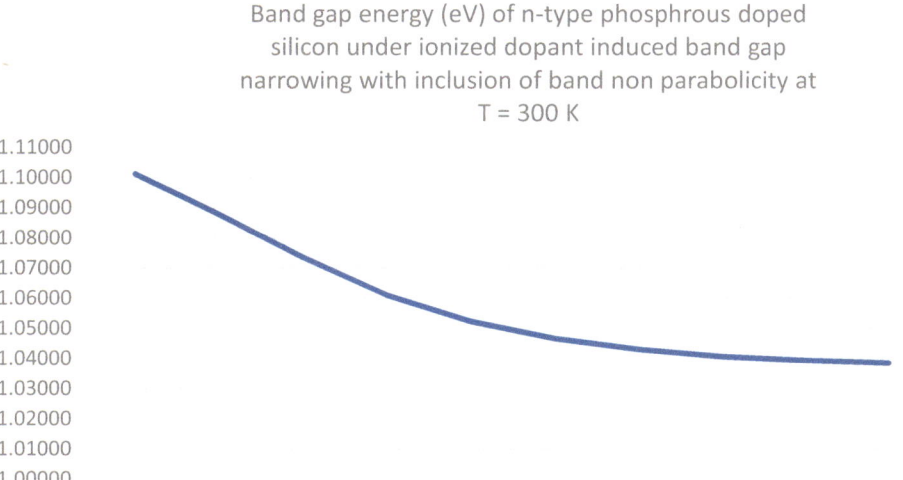

Fig. 4.12 Band gap energy (eV) E_G^n,BGN (eV) in n-type phosphorous doped silicon for substrate doping levels up to degenerate values at T = 300 K considering band non parabolicity effect on dopant ionization and dopant ionization induced band gap narrowing

Table 4.13 Actual band gap in p-type silicon at T = 300 K $E_G{}^P{}_{,BGN}$ (eV) with band non parabolicity

N_A (/cm^3)	$|\Delta E_c (N_A)|$ (eV)	$\Delta E_V (N_A)$ (eV)	$E_G{}^P{}_{,,BGN}$ (eV)
2.18×10^{17}	0.01071	9.346×10^{-3}	1.09994
1.61×10^{18}	0.01692	0.01637	1.08671
5.34×10^{18}	0.0237	0.02446	1.07184
1.1×10^{19}	0.029655	0.0316	1.058745
1.69×10^{19}	0.03343	0.03624	1.05033
2.19×10^{19}	0.03587	0.03914	1.04499
2.56×10^{19}	0.03739	0.040956	1.041654
2.81×10^{19}	0.038314	0.04206	1.039626
2.97×10^{19}	0.03887	0.04273	1.0384
3.07×10^{19}	0.03921	0.04313	1.03766

Table 4.13 lists the related parameters along with substrate doping concentrations for p-type boron doped silicon and calculated $E_G{}^P{}_{,BGN}$ (eV) values at T = 300 K.

Figure 4.13 shows the value of $E_G{}^P{}_{,BGN}$ (eV) for boron doped p-type silicon at T = 300 K with inclusion of band non parabolicity on dopant ionization and dopant ionization based band gap narrowing.

Bulk potential determination for n-type degenerately doped silicon with phosphorous at T = 300 K and also for p-type degenerately doped silicon with boron at T = 300 K under band non parabolicity effect

Bulk potential, $\varphi_{B,n}$ for n-type degenerately doped silicon with phosphorous doping at T = 300 K under band non parabolicity effect is computed for the first time through analytical modeling based derivations methods in this book by the author of this book, where this parameter is central for calculation of threshold voltage for majority carrier FET like junctionless FET and HEMT and minority carrier FET like p-MOSFET. Value of $\varphi_{B,n}$ in volt is:

$$\varphi_{B,n} = \left| \frac{E_F - E_{Fi}}{q} \right| \tag{4.14}$$

where q is not the generally substituted Coulomb charge but 1 electron charge equivalent with minus signed and E_F is in eV and E_{Fi} is intrinsic Fermi energy (eV) for n-type silicon at T = 300 K. E_F values can be gathered from Table 4.10 and E_{Fi} can be calculated from the following equation:

$$n_i^{eff} = N_c e^{\frac{(E_{Fi} - E_c)}{kT}} \tag{4.15}$$

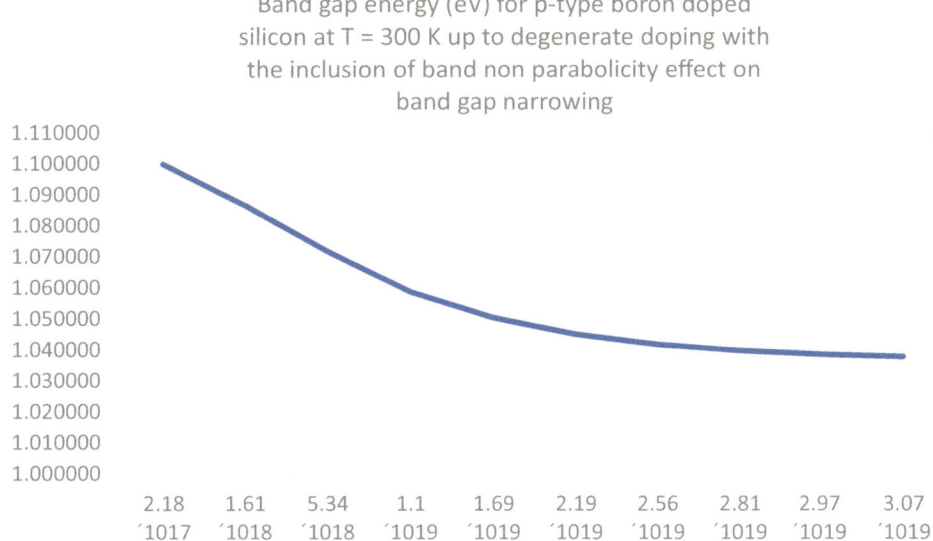

Band gap energy (eV) for p-type boron doped
silicon at T = 300 K up to degenerate doping with
the inclusion of band non parabolicity effect on
band gap narrowing

2.18	1.61	5.34	1.1	1.69	2.19	2.56	2.81	2.97	3.07
′1017	′1018	′1018	′1019	′1019	′1019	′1019	′1019	′1019	′1019

Fig. 4.13 Band gap energy (eV) $E_G{}^p{}_{,BGN}$ (eV) in p-type boron doped silicon for substrate doping levels up to degenerate values at T = 300 K considering band non parabolicity effect on dopant ionization and dopant ionization induced band gap narrowing

where E_c is impacted by both band non parabolicity induced dopant ionization and dopant ionization induced conduction band minima downshift parameter by, $E_c = 1.12 - |\Delta E_c$ $(N_D)|$ and N_c is modified from $3.23 \times 10^{19}/cm^3$ at T = 300 K for n type silicon by $3.23 \times 10^{19} \times (m_n/1.18\ m_o)^{3/2}/cm^3$. DOS values for electron effective mass m_n can be found as computed in Chap. 2 of this book and $n_i{}^{eff}$ values are transferred from Table 4.7. Now Table 4.14 lists the values of original reference substrate doping for n-type silicon, E_F, E_{Fi} and $\varphi_{B,n}$.

Figure 4.14 shows the trend of bulk potential $\varphi_{B,n}$ in n-type phosphorous doped silicon at T = 300 K including degenerate doping levels and under band non parabolicity effect:

Now for bulk potential $\varphi_{B,p}$ (V) for p-type boron doped degenerately doped silicon at T = 300 K under band non parabolicity effect on actual acceptor dopant ionization and dopant ionization induced up shift of valence band maxima $E_V = 0$ eV for silicon at T = 300 K taken as reference, the following equations detail the derivation method for the first time in this book by the author of this book:

$$\varphi_{B,p} = \frac{E_F - E_{Fi}}{q} \tag{4.16}$$

In Eq. (4.16), for p-type silicon at T = 300 K, $(E_F - E_{Fi}) < 0$ in terms of eV and q is again 1 electron equivalent charge with minus sign, so the overall $\varphi_{B,p}$ (V) is positive, which we know generally the case for positive threshold voltage of enhancement mode

Table 4.14 Bulk potential values $\varphi_{B,n}$ (V) for n-type silicon at T = 300 K with inclusion of band non parabolicity

N_D (/cm^3)	E_F (eV)	E_{Fi} (eV)	$\varphi_{B,n}$ (V)
2.18×10^{17}	0.98055	0.55365	0.4269
1.61×10^{18}	1.0216	0.55306	0.46854
5.34×10^{18}	1.0461	0.55147	0.49463
1.1×10^{19}	1.06	0.54983	0.5102
1.69×10^{19}	1.06737	0.54807	0.5193
2.19×10^{19}	1.0714	0.54675	0.52465
2.56×10^{19}	1.0736	0.54587	0.5278
2.81×10^{19}	1.07495	0.5453	0.52965
2.97×10^{19}	1.0757	0.54495	0.53075
3.07×10^{19}	1.0762	0.5447	0.5315

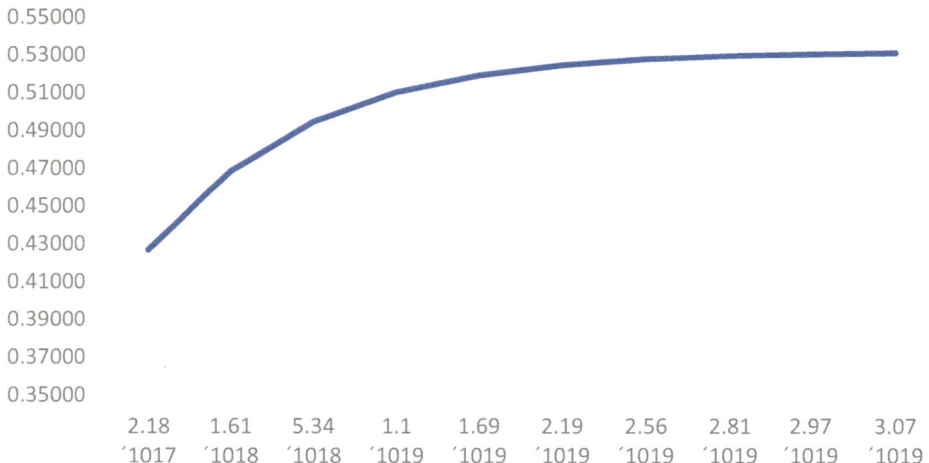

Fig. 4.14 Bulk potential $\varphi_{B,n}$ in n-type phosphorous doped silicon at T = 300 K including degenerate doping levels and under band non parabolicity effect

n-MOSFET on p-type acceptor doped silicon substrate. The other equation is:

$$n_i^{eff} = N_v e^{\frac{(\Delta E_V (N_A) - E_{Fi})}{kT}} \tag{4.17}$$

In Eq. (4.17). n_i^{eff} (/cm^3) values are from Table 4.8 of this chapter, $\Delta E_V(N_A)$ (eV) values are from Table 4.13 of this Chapter and N_V value under band non parabolicity is

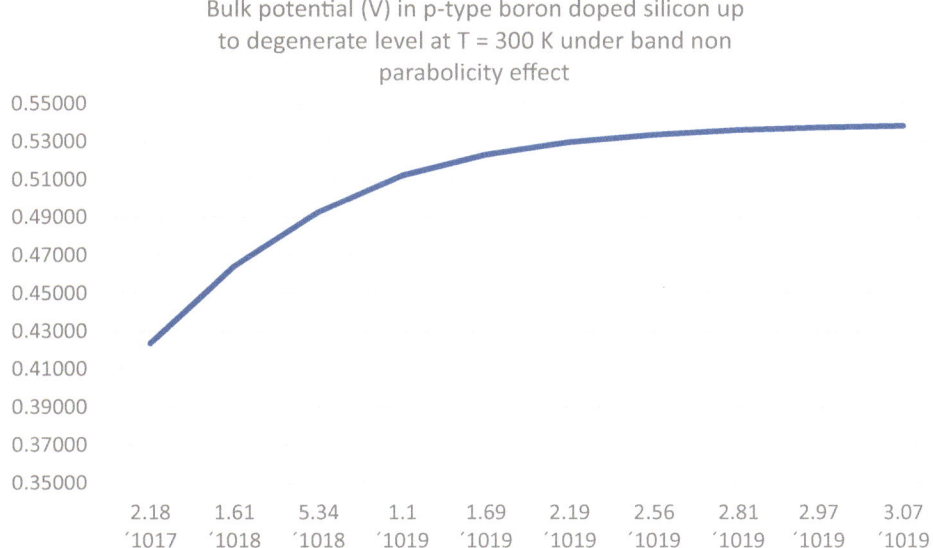

Fig. 4.15 Bulk potential $\varphi_{B,p}$ in p-type boron doped silicon at T = 300 K including degenerate doping levels and under band non parabolicity effect

1.83×10^{19} (/cm³) $\times$ $(m_p/0.81 \, m_o)^{3/2}$ where m_p is the DOS effective mass of hole under band non parabolicity and doping concentration dependent and can be found from discussion in Chap. 2 of this book. Table 4.15 lists the values of original reference substrate doping for p-type silicon, E_F, E_{Fi} and $\varphi_{B,p}$.

Figure 4.15 shows the trend of bulk potential $\varphi_{B,p}$ in p-type boron doped silicon at T = 300 K including degenerate doping levels and under band non parabolicity effect: $\varphi_{B,p}$ increases to a larger value for same bulk doping value in p-type silicon compared to $\varphi_{B,n}$ values in n-type silicon, denoting that from inversion condition under zero flat band voltage, i.e., 2 times bulk potential that largely defines the threshold voltage of MOSFET, n-MOSFET which is mostly studied and modeled in literature, has a slightly higher threshold voltage owing to its larger $\varphi_{B,p}$ compared to p-MOSFET with lower $\varphi_{B,n}$ particularly, at the higher end of substrate doping concentration values which is generally required in MOSFET to combat short channel effects based threshold voltage roll off or the off current increase. In modern FET architectures, gate integrity is increased by double gate, tri-gate and other multi-gate architectures on both n and p-FET, so bulk doping can be kept at the lower end of moderate non degenerate values.

Table 4.16 shows the ratio $\varphi_{B,p}/\varphi_{B,n}$ from which the mismatch of threshold voltage of n-MOSFET and p-MOSFET for same bulk substrate doping can be assessed as substrate doping is increased to degenerate level and band non parabolicity effect is included. Figure 4.16 shows the corresponding curve.

Table 4.15 Bulk potential values $\varphi_{B,p}$ (V) for p-type silicon at T = 300 K with inclusion of band non parabolicity

N_A (/cm^3)	E_F (eV)	E_{Fi} (eV)	$\varphi_{B,p}$ (V)
2.18×10^{17}	0.12847	0.55202	0.42355
1.61×10^{18}	0.0886	0.5528	0.4642
5.34×10^{18}	0.06192	0.5547	0.49278
1.1×10^{19}	0.04516	0.5573	0.51214
1.69×10^{19}	0.03653	0.55965	0.52312
2.19×10^{19}	0.031665	0.56133	0.52967
2.56×10^{19}	0.02883	0.56247	0.53364
2.81×10^{19}	0.02718	0.56319	0.53601
2.97×10^{19}	0.0262	0.56364	0.53744
3.07×10^{19}	0.0256	0.5638	0.5382

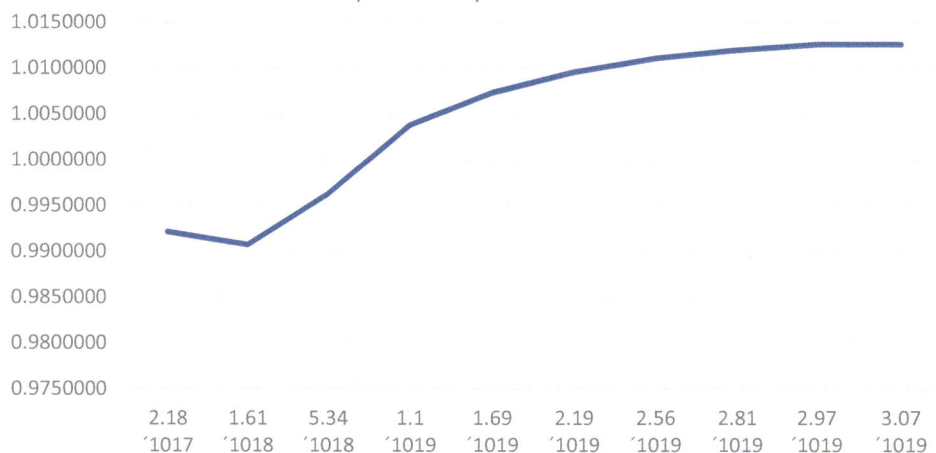

Ratio of bulk potential of p-type silicon to bulk potential of n-type silicon at T = 300 K up to degenerate substrate doping including band non parabolicity effect

Fig. 4.16 Ratio of $\varphi_{B,p}$ (V)/$\varphi_{B,n}$ (V) for the reference substrate doping up to degenerate level in silicon at T = 300 K, showing that higher $\varphi_{B,p}$ (V) compared to $\varphi_{B,n}$ (V), will reduce the gate over drive related inversion channel carrier density in n-MOSFET and reduce its drive current as with high degenerate doping, the minority electron inversion channel mobility in n-MOSFET is reduced to almost same level of minority hole inversion channel mobility in p-MOSFET, shown in minority carrier mobility computation section of this Chapter in this book

Table 4.16 The ratio of $\varphi_{B,p}/\varphi_{B,n}$ for silicon at T = 300 K with band non parabolicity

N_A or N_D (/cm³)	$\varphi_{B,p}$ (V)	$\varphi_{B,n}$ (V)	$\varphi_{B,p} / \varphi_{B,n}$
2.18×10^{17}	0.42355	0.4269	0.992153
1.61×10^{18}	0.4642	0.46854	0.990737
5.34×10^{18}	0.49278	0.49463	0.99626
1.1×10^{19}	0.51214	0.5102	1.003802
1.69×10^{19}	0.52312	0.5193	1.007356
2.19×10^{19}	0.52967	0.52465	1.009568
2.56×10^{19}	0.53364	0.5278	1.0110648
2.81×10^{19}	0.53601	0.52965	1.012008
2.97×10^{19}	0.53744	0.53075	1.0126048
3.07×10^{19}	0.5382	0.5315	1.0126058

Substrate resistivity calculation in n-type and p-type silicon at T = 300 K up to degenerate substrate doping and under band non parabolicity effect

The substrate resistivity $\rho_{maj,n}$ (ohm-cm) for majority carrier related n-type silicon at T = 300 K up to degenerate substrate doping under band non parabolicity effect is analytically modeled for the first time in this Chapter by the author of this book. The relevant equation is:

$$\rho_{maj,n} = \frac{1}{qn\mu_n^{maj}} \tag{4.18}$$

In Eq. (4.18) n is ionized majority carrier n-type phosphorous dopant at T = 300 K or N_D^+ and this parameter values can be extracted from Table 4.7 of this chapter and the values of μ_n^{maj} can be extracted from Table 4.1 of this Chapter of this book. Table 4.17 lists these parameters with computed $\rho_{maj,n}$ (ohm-cm).

Figure 4.17 shows the trend of $\rho_{maj,n}$ (ohm-cm) with regard to substrate doping including degeneracy and non parabolicity. Experimentally this value can be rapidly extracted through Hall effect based measurements but near precise analytical equation based formulation is necessary for semiconductor device professionals as p-MOSFET substrate resistivity impacts substrate current at high drain bias, a feature known as hot electron or hot hole based reliability effect and substrate resistivity is also important from the consideration of parasitic bipolar transistor induced latch up effect where the source to drain voltage remains fixated at a small nominal value but measured drain current increases substantially, eventually burning the device. For better FET performance from reliability perspective, the substrate doping must be in moderate non degenerate doping concentration but higher doping up to degenerate doping has to be avoided unless for majority carrier high speed devices like junctionless FET and HEMTs.

For majority carrier hole substrate resistivity $\rho_{maj,p}$ (ohm-cm) in p-type boron doped silicon with degenerate doping levels at T = 300 K with the inclusion of band non

Table 4.17 Computed majority carrier substrate resistivity for n-type silicon $\rho_{maj,n}$ (ohm-cm) with band non parabolicity

N_D (/cm^3)	N_D^+ (/cm^3)	μ_n^{maj} (cm^2/V-s)	$\rho_{maj,n}$ (Ω-cm)
2.18×10^{17}	2.033×10^{17}	1173.503	0.0262
1.61×10^{18}	1.3035×10^{18}	306.0725	0.01567
5.34×10^{18}	4.5635×10^{18}	148.698	9.2104×10^{-3}
1.1×10^{19}	1.0177×10^{19}	110.556	5.555×10^{-3}
1.69×10^{19}	1.623×10^{19}	96.84	3.97655×10^{-3}
2.19×10^{19}	2.1345×10^{19}	90.519	3.235×10^{-3}
2.56×10^{19}	2.511×10^{19}	87.248	2.853×10^{-3}
2.81×10^{19}	2.765×10^{19}	85.473	2.6445×10^{-3}
2.97×10^{19}	2.9269×10^{19}	84.477	2.5277×10^{-3}
3.07×10^{19}	3.0283×10^{19}	83.904	2.4598×10^{-3}

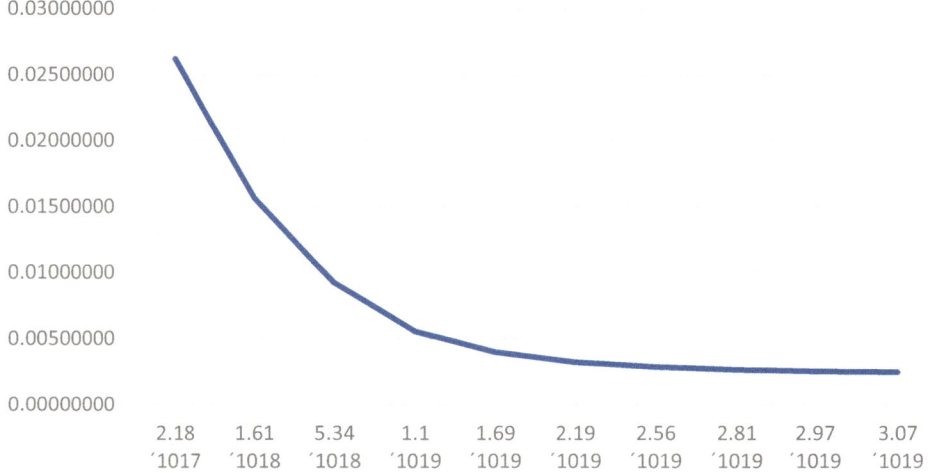

Majority carrier n-type substrate resistivity (ohm-cm) in phosphorous doped silicon up to degenerate level at T = 300 K including band non parabolicity effect

Fig. 4.17 Majority carrier n-type substrate resistivity $\rho_{maj,n}$ (ohm-cm) for phosphorous doped silicon up to degenerate doping concentrations at T = 300 K including band non parabolicity effect

Table 4.18 Computed majority carrier substrate resistivity for p-type silicon $\rho_{maj,p}$ (ohm-cm) with band non parabolicity

N_A (/cm^3)	$N_A{}^-$ (/cm^3)	$\mu_p{}^{maj}$ (cm^2/V-s)	$\rho_{maj,p}$ (ohm-cm)
2.18×10^{17}	1.8254×10^{17}	375.963	0.09107
1.61×10^{18}	1.11716×10^{18}	125.499	0.04458
5.34×10^{18}	4.23539×10^{18}	81.228	0.01817
1.1×10^{19}	1.021155×10^{19}	69.23	8.8408×10^{-3}
1.69×10^{19}	1.63859×10^{19}	64.126	5.948×10^{-3}
2.19×10^{19}	2.15169×10^{19}	61.4509	4.727×10^{-3}
2.56×10^{19}	2.52824×10^{19}	59.965	4.1225×10^{-3}
2.81×10^{19}	2.78167×10^{19}	59.125	3.8×10^{-3}
2.97×10^{19}	2.94354×10^{19}	58.643	3.6207×10^{-3}
3.07×10^{19}	3.0446×10^{19}	58.365	3.517×10^{-3}

parabolicity of valence band of p-type silicon, the following equation is invoked:

$$\rho_{maj,p} = \frac{1}{qp\mu_p^{maj}} \tag{4.19}$$

In Eq. (4.19), p is ionized acceptor density and can be replaced by values of $N_A{}^-$ (/cm^3) from Table 4.18 and $\mu_p{}^{maj}$ (cm^2/V-s) can be replaced from Table 4.4. Table 4.18 lists the related values with substrate doping concentrations for boron doped acceptors in silicon and the computed substrate resistivity $\rho_{maj,p}$ (ohm-cm) at T = 300 K.

As we can note that listed $\rho_{maj,p}$ (ohm-cm) values are larger in magnitude than $\rho_{maj,n}$ (ohm-cm) values for boron doped p-type silicon which is used as a substrate material for n-MOSFET. With degeneracy due to degenerate doping and band non parabolicity effect duly accounted for, this means substrate effects such as hot carrier generated substrate leakage current and parasitic bipolar transistor induced latch up effects will be more evident in n-type silicon substrate since its substrate resistivity is lower, as we know silicon substrate with its level of doping concentration is not as semi-insulating as GaAs which prevents these types of substrate resistivity related reliability effects in n-MOSFET and p-MOSFET. Figure 4.18 shows the trend of computed $\rho_{maj,p}$ (ohm-cm) in relation to bulk acceptor substrate doping in p-type silicon at T = 300 K with the inclusion of band non parabolicity effect:

Table 4.19 shows the relative substrate resistivity ratio of p-type boron doped substrate to n-type phosphorous doped substrate at T = 300 K.

We can see from Table 4.19 that substrate resistivity ratio for p-type silicon to n-type silicon at T = 300 K decreases as the substrate for both type becomes highly degenerate and these resistivities in magnitudes become closer and then in both types of silicon, the substrate effects related reliability effects become almost equally dominant. Figure 4.19

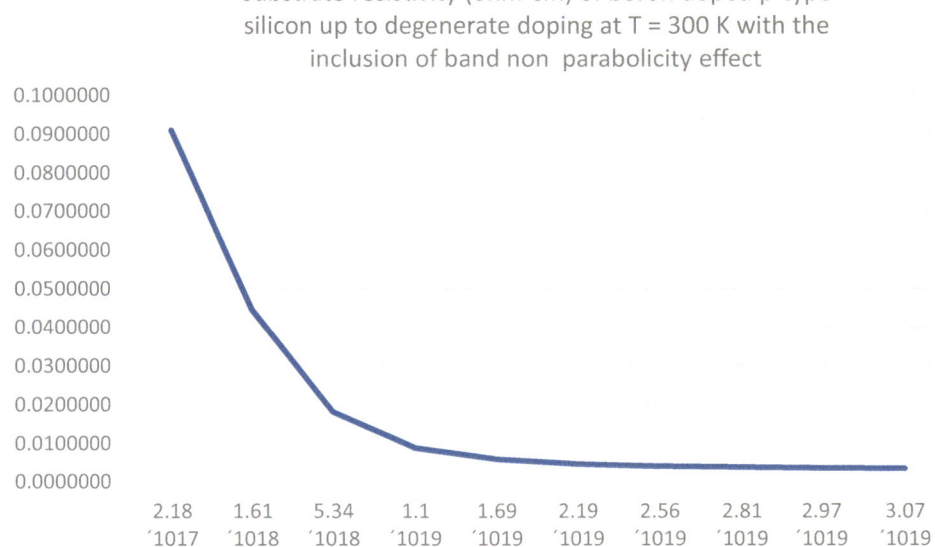

Fig. 4.18 Majority carrier p-type substrate resistivity $\rho_{maj,p}$ (ohm-cm) for boron doped silicon up to degenerate doping concentrations at T = 300 K including band non parabolicity effect

Table 4.19 The relative substrate resistivity ratio of p-type boron doped substrate to n-type phosphorous doped substrate at T = 300 K with band non parabolicity

N_A or N_D (/cm^3)	$\rho_{maj,p}$ (ohm-cm)	$\rho_{maj,n}$ (ohm-cm)	$\rho_{maj,p}/\rho_{maj,n}$
2.18×10^{17}	0.09107	0.0262	3.47595
1.61×10^{18}	0.04458	0.01567	2.84493
5.34×10^{18}	0.01817	9.2104×10^{-3}	1.97277
1.1×10^{19}	8.8408×10^{-3}	5.555×10^{-3}	1.5915
1.69×10^{19}	5.948×10^{-3}	3.97655×10^{-3}	1.49577
2.19×10^{19}	4.727×10^{-3}	3.235×10^{-3}	1.4612056
2.56×10^{19}	4.1225×10^{-3}	2.853×10^{-3}	1.44497
2.81×10^{19}	3.8×10^{-3}	2.6445×10^{-3}	1.4369446
2.97×10^{19}	3.6207×10^{-3}	2.5277×10^{-3}	1.432409
3.07×10^{19}	3.517×10^{-3}	2.4598×10^{-3}	1.42979

illustrates the trend of $\rho_{maj,p}/\rho_{maj,n}$ with respect to silicon bulk doping up to degenerate level at T = 300 K with inclusion of band non parabolicity effect. The ratio deviates more for lower doped substrate for both n-type and p-type silicon and therefore, adds benefit to p-type silicon at lower end of non degenerate substrate doping as its resistivity is quite a

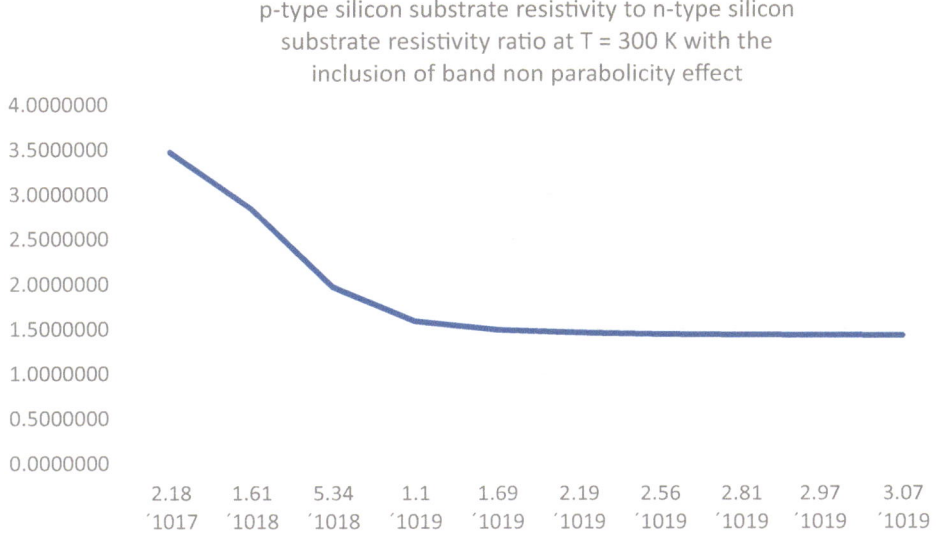

Fig. 4.19 The substrate resistivity ratio $\rho_{maj,p}/\rho_{maj,n}$ considering both boron doped p-type silicon and phosphorous doped n-type silicon with doping up to degenerate level at T = 300 K with the inclusion of band non parabolicity effect

proportion greater than unity higher than n-type silicon and will ideally track the intended purpose of FET device fabrication on semi-insulating silicon substrate.

References

1. MOSFET Performance Scaling-Part I: Historical Trends, Ali Khakifirooz and Dimitri A. Antoniadis, IEEE Transactions on Electron Devices, Vol. 55, No. 6, June 2008, pp. 1391–1400.
2. Intrinsic concentration, effective densities of states, and effective mass in silicon, Martin A. Green, Journal of Applied Physics, 67, 2944 (1990), pp. 2944–2954.
3. Reassessment of the intrinsic carrier density temperature dependence in crystalline silicon, Romain Couderc, Mohamed Amara and Mustafa Lemiti, Journal of Applied Physics, 115, 093705 (2014), pp. 093705–1 to 5.
4. Improved value for the silicon intrinsic carrier concentration from 275 to 375 K, A. B. Sproul and M. A. Green, Journal of Applied Physics, 70, 846 (1991), pp. 846–854.
5. Band gap narrowing in *n*-type and *p*-type 3C-, 2H-, 4H-, 6H-SiC, and Si, C. Persson, U. Lindefelt, and B. E. Sernelius, Journal of Applied Physics, 86, 4419 (1999), Vol. 86, Number 8, pp. 4419–4427.
6. Advanced Semiconductor Fundamentals, Robert F. Pierret. Volume VI, Second edition, Pearson Education Inc., 2003.

7. Semiconductor Device Fundamentals, Robert F. Pierret, Addison-Wesley Publication Company Inc, 1996.
8. Semiconductor Physics and Devices Basic Principles, Donald A. Neamen, Fourth Edition, McGraw Hill Companies Inc, 2012.

Some of the Very Integral Parameters Calculations for Silicon Down to Cryogenic Temperatures

Up until now, there is no analytical equation based modeling of density of states (DOS) effective mass computation for electron and hole in silicon for low temperatures down to cryogenic level such as 50 K and related actual ionization percentage calculation for phosphorous doped n-type silicon and boron doped p-type silicon at cryogenic temperatures in the range of 50 K. Accurate determinations of these parameters at cryogenic temperatures for silicon are critical from drive current perspective where channel inversion density is related to actual ionized dopants for acceptors in p-substrate n-FET and donors in n-substrate p-FET built in silicon. The author of this book in this chapter taking the temperature dependent equations for longitudinal effective mass m_l, transverse effective mass m_t, heavy hole mass m_{hh} and light hole mass m_{lh} as discussed and revealed in reference [1], derived the DOS effective masses of electron and hole down to 50 K for silicon, their actual ionization percentage in n-type and p-type silicon down to 50 K and very important conductivity effective masses of electron and hole in silicon by developments of analytically derivable equations. Once using the reference [1]. m_l, m_t, m_{hh} and m_{lh} can be computed, the previous Chapters in this book show how to derive density of states effective masses for electron from m_l and m_t and hole from m_{hh} and m_{lh} and conductivity effective masses of electron from m_l and m_t and hole from m_{hh} and m_{lh}. Chapters 2 and 3 of this book list these derivation process. We proceed first the with the derivation of DOS effective masses for electron in silicon up to 50 K in the cryogenic temperature downscaling from T = 300 K and also show the mass for temperatures up to 500 K, which can be important to assess high temperature electronics related FET performance benchmarking. In the analytical equation formulation of DOS effective mass of electron in silicon as a function of temperature, we apply three boundary conditions from three temperatures where the DOS effective mass of electron in silicon can be computed

© The Author(s), under exclusive license to Springer Nature Switzerland AG 2025   89
N. S. Ashraf, *Parameter-Centric Scaled FET Devices*, Synthesis Lectures on Emerging Engineering Technologies, https://doi.org/10.1007/978-3-031-84286-3_5

from the equations given in [1] for m_l dependence on temperature and m_t dependence on temperature.

Boundary conditions:

Temperature	DOS effective mass of electron m_n/m_o
4.2 K	1.062
100 K	1.088
300 K	1.18

Here there is a slight discrepancy in 4.2 K mn/m_o calculation from m_l and m_t values at this temperature from reference [1] when we compare the value from reference [2], so both for 4.2 K and 300 K, the listed value of m_n/m_o from reference [2] is used as the deviations are very minute when compared to [1] extracted values. This chapter is focused on cryogenic temperature modeling for dopant values in low non degenerate regime as from close to degenerate values and at degenerate values, cryogenic temperature extractions of device parameters such as effective masses and ionization percentages become very intricate and complex as band non parabolicity effect is now function of both temperature and doping value at every data point. In our analysis of the parameters illustrated in the sub-heading of chapter, the doping concentration for both n-type and p-type silicon is fixed at $10^{15}/cm^3$, so that doping dependent variation of m_n/m_o is very minute and exact temperature impact can be imposed on m_n/m_o. Now the three term polynomial equation for DOS effective masses of electron normalized to m_o, m_n/m_o as a function of temperature for silicon is given below:

$$\frac{m_n}{m_o} = 1.061 + 2.05 \times 10^{-4}T + 6.376 \times 10^{-7}T^2 \tag{5.1}$$

Table 5.1 shows the computed values of m_n/m_o for DOS effective mass of electron in silicon as a function of temperature from T = 50 K to T = 500 K.

Figure 5.1 shows the related plot of DOS effective mass of electron in silicon normalized to m_o as a function of temperature as a curve trend.

For p-type boron doped silicon with a non degenerate doping value fixed at $10^{15}/cm^3$, boundary values are computed for DOS effective mass of hole in silicon m_p normalized to m_o at T = 4.2 K, T = 100 K and T = 300 K from the m_{hh} and m_{lh} dependence related equations listed in reference [1]. Using these boundary values, the three term polynomial equation is constructed for m_p/m_o which are shown below:

Table 5.1 The computed values of m_n/m_o for DOS effective mass of electron in silicon as a function of temperature from T = 50 K to T = 500 K

T (Temperature in Kelvin)	m_n/m_o
50	1.072844
100	1.087876
150	1.106096
200	1.127504
250	1.1521
300	1.179884
350	1.210856
400	1.245016
450	1.282364
500	1.3229

DOS effective mass of electron normalized to m_o
as a function of temperature for a non degenerate
doping value of $10^{15}/cm^3$

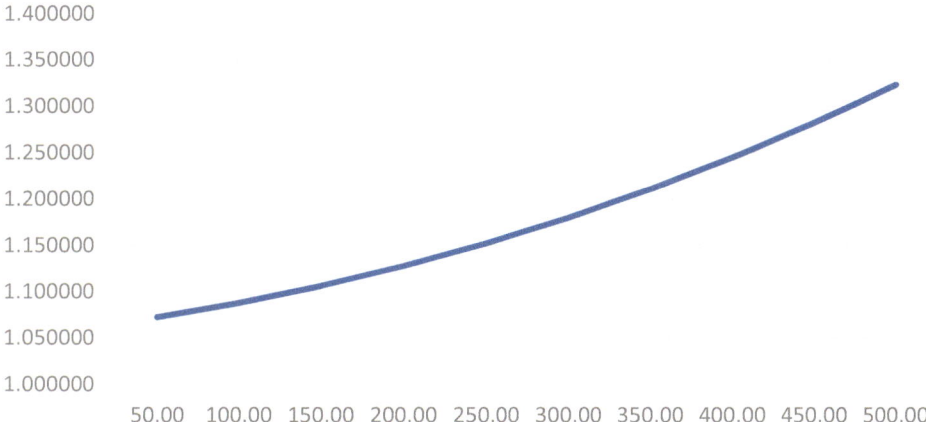

Fig. 5.1 The DOS effective mass of electron in silicon normalized to m_o as a function of temperature for a substrate non degenerate doping value of $10^{15}/cm^3$. m_n/m_o as extracted in this chapter of this book by the author for temperature as low as 50 K is very much uncommon in literature survey of reference articles

Table 5.2 The computed values of m_p/m_o for DOS effective mass of hole in silicon as a function of temperature from T = 50 K to T = 500 K

T (Temperature in Kelvin)	m_p/m_o
50	0.59825
100	0.6265
150	0.66175
200	0.704
250	0.75325
300	0.8095
350	0.87275
400	0.943
450	1.02025
500	1.1045

Boundary conditions:

Temperature	DOS effective mass of electron m_n/m_o
4.2 K	0.579
100 K	0.627
300 K	0.81

$$\frac{m_p}{m_o} = 0.577 + 3.55 \times 10^{=4}T + 1.4 \times 10^{-6}T^2 \qquad (5.2)$$

Table 5.2 shows the computed values of m_p/m_o for DOS effective mass of hole in silicon as a function of temperature from T = 50 K to T = 500 K.

Figure 5.2 shows the related plot of DOS effective mass of hole in silicon normalized to m_o as a function of temperature as a curve trend for a fixed doping concentration of $10^{15}/cm^3$ where doping dependence of m_p is negligible considering its base setting at T = 300 K is 0.81 m_o [2, 3].

For conductivity effective mass of electron in silicon as a function of temperature, variation equations for m_l and m_t can be found in [1]. The derivation process of conductivity effective mass m_{cn} for silicon is discussed in Chap. 3 of this book. For deriving the three term analytical equations of m_{cn} normalized to m_o as a function of substrate temperature T in kelvin, we compute the following boundary conditions:

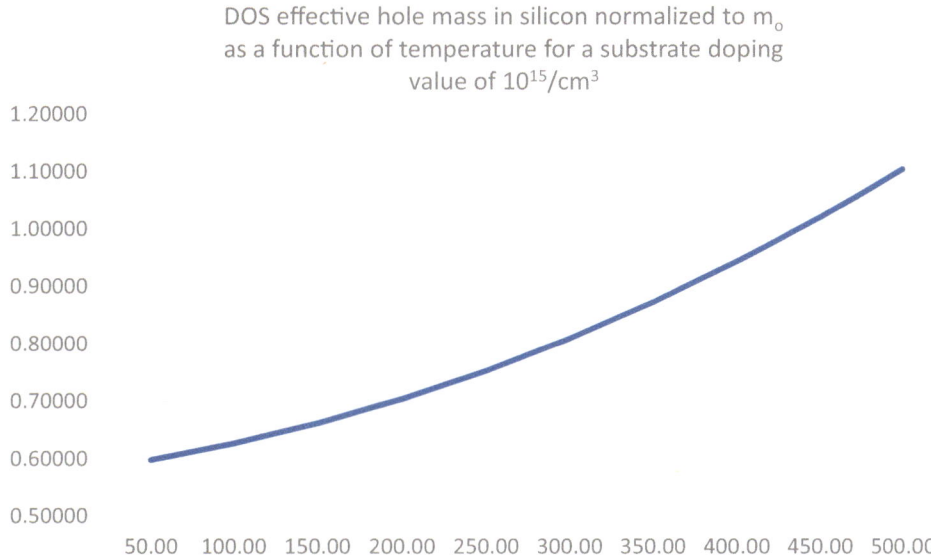

Fig. 5.2 The DOS effective mass of hole in silicon normalized to m_0 as a function of temperature for a substrate non degenerate doping value of $10^{15}/cm^3$. m_p/m_0 as extracted in this chapter of this book by the author for temperature as low as 50 K is very much uncommon in literature survey of reference articles

Boundary conditions:

Temperature	Conductivity effective mass of electron m_{cn}/m_0
4.2 K	0.259
100 K	0.268
300 K	0.298

$$\frac{m_{cn}}{m_o} = 0.259 + 7.424 \times 10^{=5}T + 1.894 \times 10^{-7}T^2 \qquad (5.3)$$

Table 5.3 The values of m_{cn} normalized to m_o for the set of substrate temperatures down to cryogenic level

T (Temperature in Kelvin)	m_{cp}/m_o
50	0.263186
100	0.268318
150	0.274398
200	0.281424
250	0.289398
300	0.298318
350	0.3081855
400	0.319
450	0.3307615
500	0.34347

Table 5.3 shows the values of m_{cn} normalized to m_o for the set of substrate temperatures down to cryogenic level. These computations and Eq. (5.3) are the first of kinds derivations as narrated in this book by the author for the first time.

Figure 5.3 shows the related plot of conductivity effective mass of electron in silicon normalized to m_o as a function of temperature as a curve trend for a fixed doping concentration of $10^{15}/cm^3$ where doping dependence of m_{cn} is negligible considering its base setting at T = 300 K is 0.298 m_o [3.4].

For conductivity effective mass of hole in silicon as a function of temperature, variation equations for m_{lh} and m_{hh} can be found in [1]. The derivation process of conductivity effective mass m_{cp} for silicon is discussed in Chap. 3 of this book. For deriving the three term analytical equations of m_{cp} normalized to m_o as a function of substrate temperature T in kelvin, we compute the following boundary conditions:

Boundary conditions:

Temperature	Conductivity effective mass of hole m_{cp}/m_o
4.2 K	0.396
100 K	0.423
300 K	0.544

$$\frac{m_{cp}}{m_o} = 0.395 + 1.678 \times 10^{=4}T + 1.093 \times 10^{-6}T^2 \tag{5.4}$$

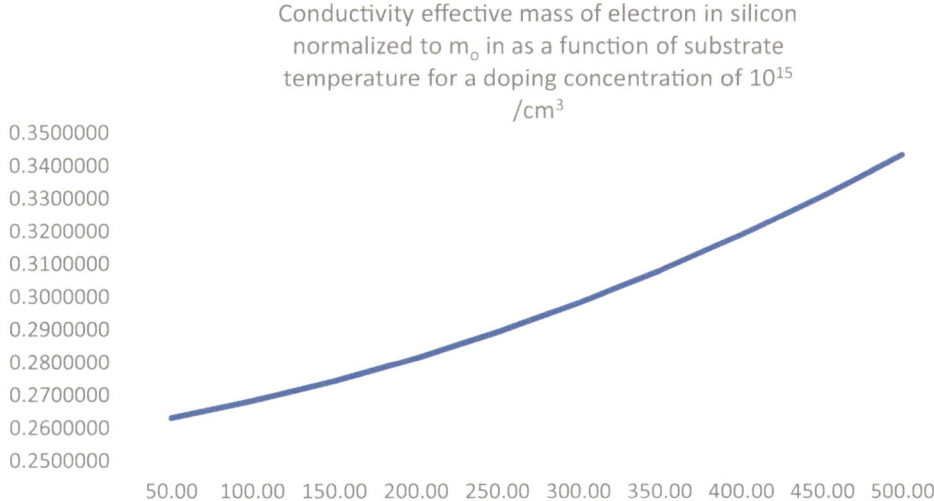

Fig. 5.3 The conductivity effective mass of electron in silicon normalized to m_o as a function of temperature for a substrate non degenerate doping value of $10^{15}/cm^3$. m_{cn}/m_o as extracted in this chapter this book by the author for temperature as low as 50 K is very much uncommon in literature survey of reference articles

Table 5.4 The values of m_{cp} normalized to m_o for the set of substrate temperatures down to cryogenic level

T (Temperature in Kelvin)	m_{cp}/m_o
50	0.4061225
100	0.42271
150	0.4447625
200	0.47228
250	0.5052625
300	0.54371
350	0.5876225
400	0.637
450	0.6918425
500	0.75215

Table 5.4 shows the values of m_{cp} normalized to m_o for the set of substrate temperatures down to cryogenic level. These computations and Eq. (5.4) are the first of kinds derivations as narrated in this book by the author for the first time.

Figure 5.4 shows the related plot of conductivity effective mass of hole in silicon normalized to m_o as a function of temperature as a curve trend for a fixed doping concentration of $10^{15}/cm^3$ where doping dependence of m_{cp} is negligible considering its base

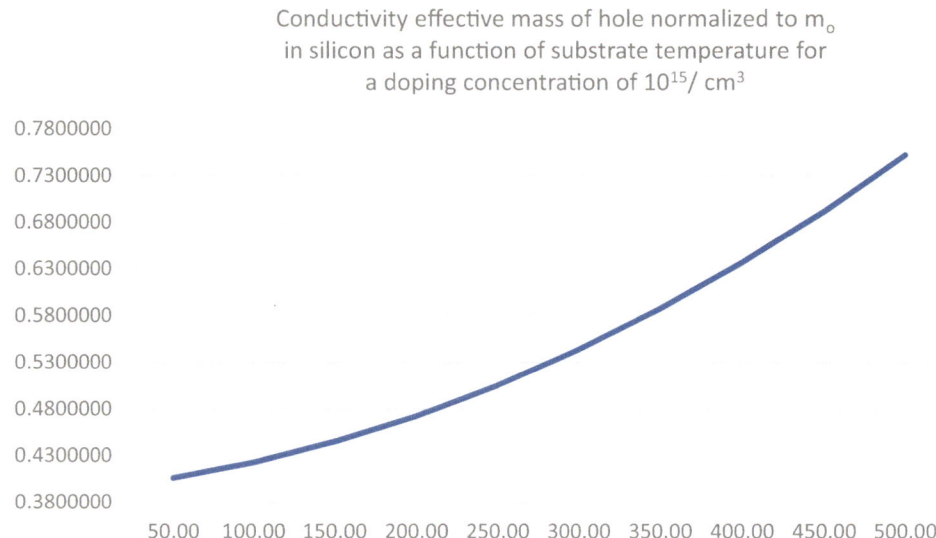

Conductivity effective mass of hole normalized to m_o in silicon as a function of substrate temperature for a doping concentration of $10^{15}/cm^3$

Fig. 5.4 The conductivity effective mass of hole in silicon normalized to m_0 as a function of temperature for a substrate non degenerate doping value of $10^{15}/cm^3$. m_{cp}/m_0 as extracted in this chapter of this book by the author for temperature as low as 50 K is very much uncommon in literature survey of reference articles

setting at T = 300 K is 0.544 m_0 [3.4]. Comparing Figs. 5.3 and 5.4, it is evident that temperature impact from cryogenic level to temperatures above than 300 K, is more on m_{cp} variation compared to m_{cn} variation. So the majority carrier p-HEMT and junctionless p-FET and minority carrier p-FET (inversion holes with m_{cp} conductivity effective mass) will be relatively more sensitive to cryogenic temperatures and at high temperatures due to high m_{cp} variation when compared to transport level drain current, mobility and saturated drift velocity assessment for majority carrier n-HEMT and junctionless n-FET and minority carrier n-FET (inversion electrons with m_{cn} conductivity effective mass).

Computations of actually ionized percentage of dopants in n-type phosphorous doped silicon and p-type boron doped silicon as a function of substrate temperature down to cryogenic level

For n-type phosphorous doped silicon, for calculation of actual ionized donor dopant percentage as a function of substrate temperature down to cryogenic conditions at 50 K and up to some high substrate temperature 400 K, modification of Eq. 1.3 of Chap. 1 of this book will be required. For sufficiently non degenerate doping concentration of $10^{15}/cm^3$ which will remain fixed for calculation of ionized donor percentage as a function of substrate temperature, parameter b value from Eq. 1.1 of Chap. 1 of this book is 0.99999 and E_{dop} from Eq. 1.2 is 0.045499 eV for $10^{15}/cm^3$ substrate doping concentration. Equation 1.3 of Chap. 1 of this book is modified as:

Table 5.5 The donor ionized factor and ionization percentage for n-type phosphorous doped silicon as a function of substrate temperature for silicon

T in Kelvin	$N_D^+/(10^{15}/cm^3)$	$N_D^+/(10^{15}/cm^3)$ %
50	0.1459517	14.59517
100	0.9375425	93.75425
150	0.993585	99.3585
200	0.9983	99.83
250	0.9995	99.95
300	1.0000	100.00
400	1.0000	100.00

$$n_1(T) = 3.23 \times 10^{19} \left(\frac{m_n(T)}{1.18m_o} \right)^{3/2} \left(\frac{T}{300} \right)^{3/2} e^{-\frac{E_{dop}}{kT}} \tag{5.5}$$

m_n (T) values can be found from Table 5.1. In Eq. (5.5) E_{dop} approximates $E_c - E_D$ in eV and as E_c shifts in energy, E_D also shifts equal amount as E_{dop} as seen from Eq. 1.2 of first chapter of this book is only shifted with more substrate doping concentration numerical value. Now Eq. (5.5) has to be used in conjunction with Eqs. 1.4 and 1.5 of Chap. 1 of this book to arrive at final $N_D^+/(10^{15}/cm^3)$ factor and eventually percentage.

Table 5.5 lists the donor ionized factor and ionization percentage for n-type phosphorous doped silicon as a function of substrate temperature for silicon. These computations are for the first time reported in this chapter of this book.

Figure 5.5 illustrates the ionized percentage of dopants in phosphorous doped n-type silicon as a function of substrate temperature down to cryogenic level for a sufficiently low non degenerate substrate doping concentration of $10^{15}/cm^3$.

For p-type boron doped silicon, for calculation of actual ionized acceptor dopant percentage as a function of substrate temperature down to cryogenic conditions at 50 K and up to some high substrate temperature 400 K, modification of Eq. 1.8 of Chap. 1 of this book will be required. For sufficiently non degenerate doping concentration of $10^{15}/cm^3$ which will remain fixed for calculation of ionized acceptor percentage as a function of substrate temperature, parameter b value from Eq. 1.6 of Chap. 1 of this book is 0.99999 and E_{dop} from Eq. 1.7 is 0.044388 eV for $10^{15}/cm^3$ substrate doping concentration. Equation 1.8 of Chap. 1 of this book is modified as:

$$n_1(T) = 1.83 \times 10^{19} \left(\frac{m_p(T)}{0.81m_o} \right)^{3/2} \left(\frac{T}{300} \right)^{3/2} e^{-\frac{E_{dop}}{kT}} \tag{5.6}$$

m_p (T) values can be found from Table 5.2. In Eq. (5.6) E_{dop} approximates $E_D - E_V$ in eV and as E_V is set reference in energy as 0 eV, due to band gap variation with temperature as E_c shifts, E_D also shifts equal amount as E_{dop} as seen from Eq. 1.7 of first chapter of this book is only shifted with more substrate doping concentration numerical value. Now

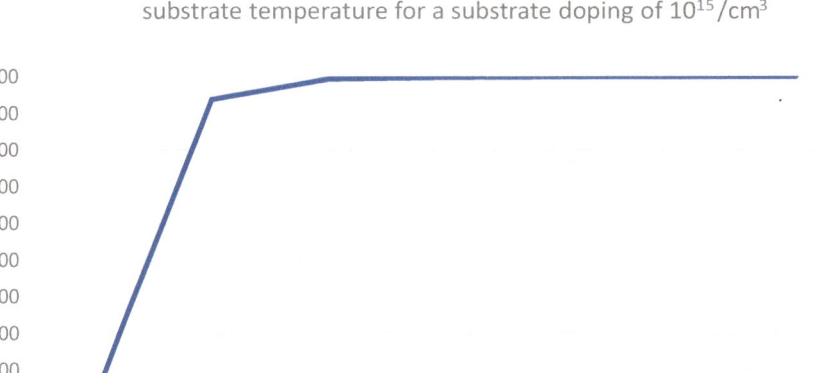

Actual ionized donor dopant percentage in phosphorous doped n-type silicon as a function of substrate temperature for a substrate doping of $10^{15}/cm^3$

Fig. 5.5 The ionized percentage of dopants in phosphorous doped n-type silicon as a function of substrate temperature down to cryogenic level for a sufficiently low non degenerate substrate doping concentration of $10^{15}/cm^3$

Table 5.6 The acceptor ionized factor and ionization percentage for p-type boron doped silicon as a function of substrate temperature for silicon	T in Kelvin	$N_A^-/(10^{15}/cm^3)$	$N_A^-/(10^{15}/cm^3)$ %
	50	0,078,467	7.8467
	100	0.81121	81.121
	150	0.975385	97.5385
	200	0.9935	99.35
	250	0.9975	99.75
	300	0.9988	99.88
	400	0.9996	99.96

Eq. (5.6) has to be used in conjunction with Eq. (1.9) and (1.10) of Chap. 1 of this book to arrive at final $N_A^-/(10^{15}/cm^3)$ factor and eventually percentage.

Table 5.6 lists the acceptor ionized factor and ionization percentage for p-type boron doped silicon as a function of substrate temperature for silicon. These computations are for the first time reported in this book.

Figure 5.6 illustrates the ionized percentage of dopants in boron doped p-type silicon as a function of substrate temperature down to cryogenic level for a sufficiently low non degenerate substrate doping concentration of $10^{15}/cm^3$.

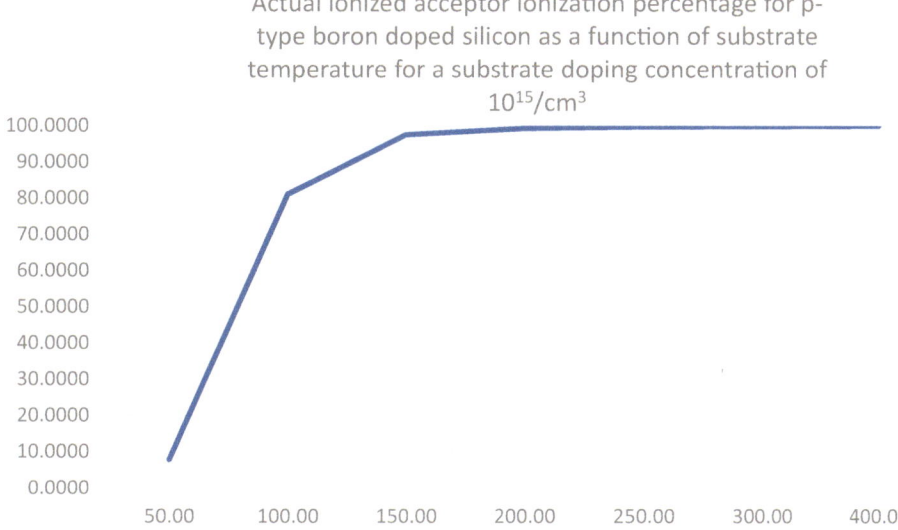

Fig. 5.6 The ionized percentage of dopants in boron doped p-type silicon as a function of substrate temperature down to cryogenic level for a sufficiently low non degenerate substrate doping concentration of $10^{15}/cm^3$

For acceptor doped p-silicon with boron, cryogenic temperatures low as 50 K and lower affects the ionized percentage with higher reduction compared to n-type phosphorous doped silicon. So n-FET operating at cryogenic temperature will have higher percentage of neutral dopants at cryogenic temperature and near threshold or subthreshold with low drain voltage, n-FET because of these relatively lower ionization of acceptor dopants in p-substrate, will suffer from lower mobility, although due to lower percentage of ionization, threshold voltage notably being higher in n-FET and p-FET due to low temperature level operation (increase of band gap), will be slightly lower in n-FET than in p-FET due to bulk potential dependence on ionized dopants and in n-FET, this bulk potential at cryogenic temperature will be slightly lower owing to lower ionized acceptor carriers. This may give more advantage to n-FET for gate overdrive based inversion carrier density enhancement at cryogenic temperature and with higher drain voltage, ballistic operation is more visible in n-FET at cryogenic temperature down to 4.2 K when drain voltage is sufficiently high and long array of neutral dopants thus increase the mean time between scattering significantly in n-FET than in p-FET where the n-substrate such as phosphorous has the neutral doping density at cryogenic temperature being lower than the same in n-FET.

Table 5.7 lists the $N_A{}^-/N_D{}^+$ as a function of substrate temperature in silicon showing how this ratio of ionized carriers is impacted near cryogenic temperature down to 50 K.

Table 5.7 Listing of N_A^-/N_D^+ as a function of substrate temperature in silicon

T in Kelvin	N_D^+ (/cm^3)	N_A^- (/cm^3)	N_A^-/N_D^+
50	1.459517×10^{14}	7.84671×10^{13}	0.53762
100	9.375425×10^{14}	8.1121×10^{14}	0.86525
150	9.93585×10^{14}	9.75385×10^{14}	0.98168
200	9.983×10^{14}	9.935×10^{14}	0.995192
250	9.995×10^{14}	9.975×10^{14}	0.998
300	10^{15}	9.988×10^{14}	0.9988
400	10^{15}	9.996×10^{14}	0.9996

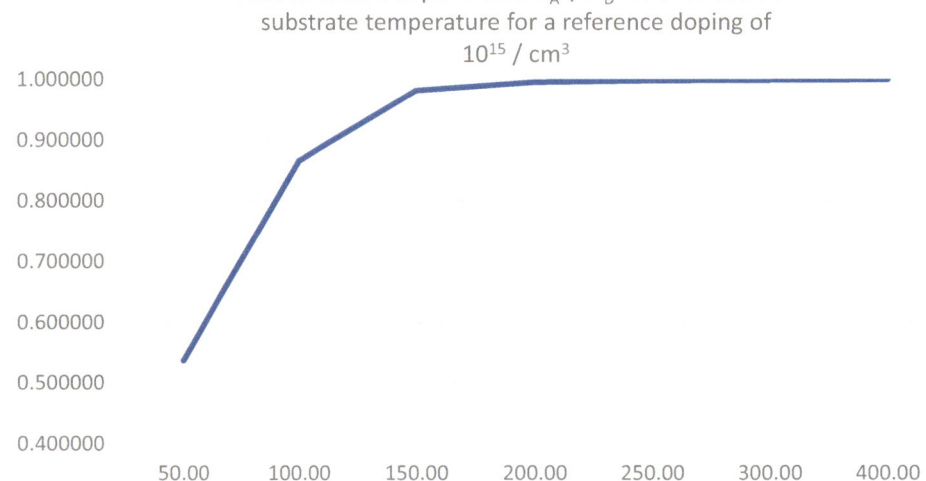

Fig. 5.7 The relative ratio N_A^-/N_D^+ as a function of substrate temperature down to cryogenic level for a reference substrate doping of 10^{15}/cm^3. At temperatures near 50 K and below, the ratio is severely attenuated and becomes slightly higher than 0.5 at 50 K. Also, at higher temperatures up to 400 K, the acceptor ionization is still not 100%, so the ratio is below 1 at T = 400 K

Figure 5.7 shows the relative ratio N_A^-/N_D^+ as a function of substrate temperature down to cryogenic level for a reference substrate doping of 10^{15}/cm^3.

Neutral Impurity Scattering Effects on Carrier Transport Related Mobility of Silicon FET at Cryogenic Temperature Operation No Error Detected

At cryogenic temperature, the peak mobility for carrier transport for inversion layer channel carriers in n-FET and p-FET is determined by phonon limited mobility with temperature dependence of $(T)^{-3/2}$, so it is obvious that peak mobility at cryogenic temperature operation is much higher than mobility at $T = 300$ K or room temperature operation. But, at cryogenic temperature, due to intense incomplete ionization when the substrate doping is towards the lower end of non degenerate doping concentration, there is considerable concentration of neutral dopants in the depletion region and in the transport physics analysis, thus neutral impurity scattering of carriers at various phases of gate and drain voltage bias, in addition to generally computed ionized impurity scattering is needed to be modeled and incorporated to determine the effective minority carrier scattering time between collision that determines the channel mobility from the Drude mobility equation discussed in Chaps. 3 and 4. In this sub-section we will probe over the role of neutral impurity scattering on minority carrier transport first for n-FET for subthreshold bias, near threshold bias and inversion level bias. A neutral acceptor dopant in the depletion region can be visualized as a negatively ionized acceptor bound with a hole (which is not free moving or mobile). When in n-FET, the electron comes in the vicinity of this neutral acceptor atom, the first in the array of many of those in the first few layers of depletion region width near cryogenic temperature operation, the natural lift-up force or repulsive force that emanates from negatively charged acceptor exclusively, now gets modified as there is a bound hole attached to the negatively charged acceptor making is charge-neutral and hence, the electron encounters an attractive force towards this bound hole of neutral acceptor, resulting in lowering of repulsive force as the electron carrier passes over neutral atom in the vicinity of dopant's scattering cross-section. This results in loss of momentum and initial mobility decrease from the first scattering encounter with the first neutral dopant in the series, the condition is device is biased in subthreshold region, the gate voltage is sufficiently low than threshold voltage causing much less inversion channel carrier concentration and the forward drift field is also low as there is very small drain voltage. A neutral dopant atom positioned in the depletion region next to another neutral dopant atom, possesses equipotential voltage and zero forward electric field and that can extend over more neutral dopants in the array positioned in the depletion region. So, when the repulsive force is decreased and the carrier is driven toward the neutral dopant by it bound hole, while it passes the first dopant with a probabilistic collision, carrier's forward momentum is decreased and for successive neutral dopant atoms, this continuous decrease of repulsive force-field around the carrier, makes it stuck to move in the forward direction as equipotential state condition negates any sufficient forward drift force generation for these carriers when the device is biased in the subthreshold region. This is the main reason of subthreshold mobility decrease even though time between scattering is increased, where scattering is meant for ionized dopants which may be sparsely situated in the depletion region. This condition and physical analysis are not properly explained or accounted for in the reference [4] for T = 77 K experimentally determined inversion layer mobility for different substrate doping concentrations as a function of vertical gate electric field and low drain bias.. The original notion of scattering time elongation at cryogenic temperature is a misnomer as in the subthreshold region bias to near threshold region bias, due to significant neutral impurity scattering and zero net forward drift field due to equipotential force-field condition taking all the neutral dopants together and after the first momentum loss, mean momentum loss fluctuates over this value with scattering probability with succeeding neutral dopants. This situation drastically changes, while the gate voltage is inversion level bias condition with sufficient gate overdrive owing to mean threshold voltage decrease due to lower concentration of ionized dopants as device is operated near cryogenic temperature, drain voltage is also set higher for high enough lateral drift field, now due to equipotential voltage distribution over the neutral dopants array, the drain voltage bias quickly attracts the carriers with low momentum with negligible channel voltage drop towards the source, This raises the forward momentum of the carriers and now through the effect of time between collision being elongated, the carriers get sufficiently high drift velocity to reach toward the drain and contribute to higher on current. If the channel length is smaller than the length between successive collision of the carriers in n-FET or p-FET at cryogenic temperature, ballistic transport is possible where through non equilibrium transport, the carrier temperature is not raised but their drift velocity is almost 4–5 times higher than thermal velocity silicon at T = 300 K, which is an indicator of maximum saturated drift velocity of inversion carriers in silicon n-FET and p-FET at T = 300 K.

The physics of neutral impurity scattering on majority carrier transport such as junctionless n-FET and n-HEMT devices is different than what has been described for minority carrier transport based n-FET and p-FET. In majority carrier FET where electron is majority carrier, substrate is donor doped and a neutral donor has a positively charged donor dopant with a bound electron which is thus not mobile. When these neutral donor dopants are arrayed in the depletion region of junctionless n-FET or in n-HEMT, the carrier electron in the vicinity of the first neutral donor dopant, now encounters a repulsive force by the bound electron of the neutral donor dopant, resulting a lift-up force that increases the momentum of the carrier electron slightly through momentum redistribution in the probability of a collision. So, the electron's mobility in this case by the first interaction of neutral donor dopant is increased, again for succeeding donor dopants, since there is equipotential voltage between the neutral dopants in the array, this mean momentum of the carriers remain low and slightly changed by the increasing repulsive effect of the adjacent bound electron of every succeeding neutral donor dopant. Here also absence of net forward force or field, decreases the subthreshold mobility or drift velocity like in n-FET or p-FET discussed earlier, but the subthreshold mobility for this majority carrier FET like device such as junctionless n-FET or n-HEMT will be higher than the n-FET or p-FET of minority carrier transport based devices. During on condition or high gate and drain bias, majority carriers in a repulsive scattering setting, will move with higher mobility and velocity with high enough drain voltage where the on current for sufficiently low non degenerate substrate doping, can be slightly higher than n-FET or p-FET. The cryogenic temperature operation is also needed for ultra low power consumption purposes where the subthreshold leakage current or I_{off} is orders of magnitude lower than I_{on} or the device on current under inversion and saturation bias condition. For n-FET and p-FET, owing to much lower subthreshold mobility and drift velocity, the I_{off} is generally a tenth order or 2 lower than the I_{off} for majority carrier n-FET like junctionless FET or n-HEMTs. We may hypothetically form the overall scattering time taking the case of ionized scattering and neutral impurity scattering for minority carrier n-FET as:

$$\tau_{eff} = \frac{\tau_{ionized} \tau_{neutral}}{(\tau_{ionized} + \tau_{neutral})} \tag{5.7}$$

$$\tau_{eff} = \frac{\tau_{neutral}}{1 + \frac{\tau_{neutral}}{\tau_{ionized}}} \tag{5.8}$$

From Eq. (5.8) we see that even though for higher concentration of neutral atoms at cryogenic temperatures, the numerator in the Eq. (5.8) can be larger due to mean time between scattering increase, in the denominator, the ratio factor term can be larger also, hence overall denominator can be of higher value tending to reduce the overall τ_{eff} in the subthreshold bias condition and near threshold bias condition before in the inversion regime operation, the numerator factor in the Eq. (5.8) dominates over denominator such that τ_{eff} is now increased at much lower cryogenic temperature down to 4.2 K (Figs. 5.8 and 5.9).

Neutral impurity scattering for non-degenerate doping concentrations previously unaccounted at T = 300 K, impact of surface roughness scattering at cryogenic temperatures and ballistic transport at cryogenic temperatures

Chapters 1 and 2 of this book written by the author of this book showed through combinations of analytical equations described in reference [5] that between higher end of $10^{17}/cm^3$ to some higher value of $10^{19}/cm^3$ dopants are not 100% ionized for silicon at T = 300 K and near $10^{17}/cm^3$ to $10^{19}/cm^3$ the ionized concentration first drops to the level of between 90% and less and then for n-type phosphorous doped silicon at T = 300 K, the doping goes to a minima near 80% and for boron doped p-type silicon near 69% at T = 300 K illustrated by tables and plots in Chaps. 1 and 2 of this book. It means from some $10^{17}/cm^3$ towards $10^{19}/cm^3$ when the doping regime is between non-degenerate to small degenerate level, this incomplete ionization in silicon for both n-type and p-type dopants at T = 300 K suggest, there are some good concentrations of neutral impurity dopants in these concentrations range in silicon and as a result, for subthreshold and near threshold operation, where the inversion charge density is very low and screening

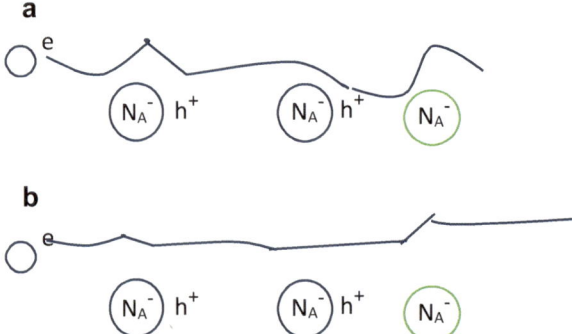

Fig. 5.8 **a** Electron travelling collision path through neutral acceptor atoms and then over an ionized acceptor atom at subthreshold bias (very low inversion carrier density and very low drain bias or drift field). Initially repelled and attracted by the bound hole, the electron momentum decreases and it continues in the collision path with very low mean transport trajectory and then near the ionized acceptor, the field repels or lifts the electron up in transport. **b** Electron travelling collision path through neutral acceptor atoms and then over an ionized acceptor atom at inversion regime bias (high inversion carrier density and high drain bias or high drift field). Initially repelled and attracted by the bound hole with screening, the electron momentum decreases and it continues in the collision path with energetic transport trajectory governed by the high lateral field and then near the ionized acceptor, the field repels or lifts the electron up in transport in high momentum and energy ballistic mode

of dopants by inversion layers is negligible, these neutral dopants will reduce subthreshold mobility in addition to ionized impurity scattering and near threshold mobility by the scattering related transport in presence of arrays of neutral dopants and ionized dopants in the depletion region as discussed in the foregoing analysis and this consideration was not addressed in the widely cited reference [4].

Reference [4] showed that for degenerate doping at its extreme, the threshold voltage is higher for T = 300 K due to bulk potential increase but that is also for the case when there is complete ionization of dopants beyond some 10^{19}/cm^3 substrate doping, only discussed in [5] and through substantiated analysis in Chaps. 1 and 2 of this book by the author but the authors of the reference [4] did not analyze this high threshold voltage's connection to the requirement of higher vertical field through high gate voltage and due to high surface band bending, surface roughness scattering becomes the principal limiting factor of inversion channel mobility both at low drain voltage and high drain voltage. Reference [4] also showed that at T = 77 K, near degenerate substrate doping, the interface roughness scattering again becomes the main mobility reduction factor for inversion channel mobility, but again this happens for near 100% ionization of dopants for both n-type and p-type silicon as discussed through computations and tables and figures in this chapter by the author. Due to this near 100% ionization of substrate dopant concentrations through the temperature related adjusted terms from the reference [5], again

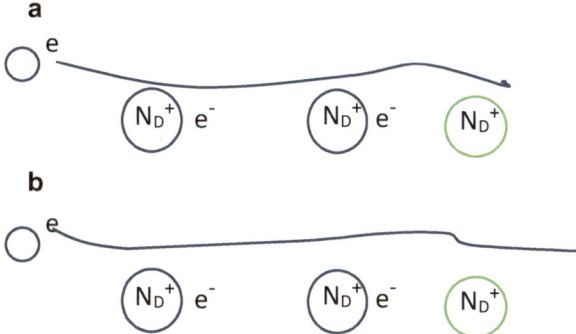

Fig. 5.9 a Electron travelling collision path through neutral donor atoms and then over an ionized donor atom at subthreshold bias (very low majority carrier density and very low drain bias or drift field). Initially attracted and repelled by the bound electron, the electron carrier momentum increases and it continues in the collision path with very low mean transport trajectory and then near the ionized donor, the field attracts or directs the electron toward the dopant in transport. **b** Electron travelling collision path through neutral donor atoms and then over an ionized donor atom at inversion bias regime (high majority carrier density and high drain bias or drift field). Initially attracted and repelled by the bound electron, the electron carrier momentum increases and it continues in the collision path with energetic transport trajectory governed by high lateral field and then near the ionized donor, the field attracts or directs the electron toward the dopant in transport but electron momentum and energy are high suitable for ballistic transport

the threshold voltage due to this vast ionized doping concentration even near cryogenic temperature in silicon, at high vertical field through the requirements of high gate voltage, develops high surface band bending and inversion layer is now subject to strong surface roughness scattering that decreases the phonon related peak mobility gain at cryogenic temperature to room temperature minimum inversion layer mobility level shown in [4] but not explained properly. If the voltage scaling limit consequential to technology node reduction is considered, due to higher threshold voltage shift due to wider band gap in silicon at cryogenic temperature operation of n-FET and p-FET, the gate overdrive is very low and therefore, the inversion layer faces boundary layer scattering in gate all around nanowire FETs in addition to increased surface roughness scattering when the doping is close to degenerate values.

Both boundary layer scattering and interface roughness scattering at cryogenic temperature operation of n-FET and p-FET can be compensated by sufficient intrinsic level non degenerate doping of the substrate near $10^{14}/cm^3$ and improving gate to channel integrity through gate all around structure, nanosheet and complementary FET structure at advanced node. At very low non degenerate substrate doping, phonon related peak mobility gain is close to 100% as temperatures are reduced to cryogenic level and as has been explained in the foregoing analysis that a long array of neutral dopants that results out of more intense incompletely ionized dopants for low non degenerate substrate doping concentrations and with higher drain voltage applied, can trigger ballistic transport in FET as the source side electrons get speedily attracted to the drain voltage penetration in the channel to the source side through the equipotential region of the neutral dopants in the depletion region, the mean time between scattering increases considerable factors more than the gate node dimension approaching 2 to 1 nm of n-FET and p-FET, the electrons travel with high kinetic energy synonymous with ballistic transport that raises the drift velocity 4–5 times more than thermally limited drift velocity of silicon at T = 300 K. So, operating the n-FET and p-FET with improved gate to channel integrity and very low non degenerate substrate doping, enablement of ballistic transport increases the drive current and reduces the subthreshold current further by neutral dopants induced reduction of subthreshold mobility in minority carrier transport in n-FET and p-FET and the desired I_{on}/I_{off} ratio > 10^6 can be ingeniously obtained at cryogenic temperature near T = 4.2 K operation of these devices.

Preview of Chaps. 6 and 7: in Chaps. 1–this chapter in this book, the author of this book provided novel insightful device physics based analysis and discussion of several crucial parameters of silicon with novel illustrative Tables and Figures to show how these device parameters are central for n-FET and p-FET performance analysis at T = 300 K down to cryogenic temperatures. In Chap. 6, there is no novel table or plot based analysis, rather highly informative reference based review discussion of gate all around (GAA) nanowire FET transport and performance assessment from device physical overview. In Chap. 7, the author of this book extends this review for nanosheet n-FET architectures which has become norm for 2–3 nm device node as per Moore's Law based roadmap. The goal is to analyze the best reference articles for Chaps. 6 and 7 and also illuminate with this book's author's distinctive and salient observations that may be additionally deduced from the contents of these reference articles chosen for Chaps. 6 and 7.

References

1. Low Temperature Electronics, Physics, Devices, Circuits and Applications, Edmundo A. Gutiérrez-D., M. Jamal Deen and C. Claeys, Academic Press, 2001.
2. Advanced Semiconductor Fundamentals, Robert F. Pierret. Volume VI, Second edition, Pearson Education Inc., 2003.
3. Semiconductor Device Fundamentals, Robert F. Pierret, Addison-Wesley Publication Company Inc, 1996.
4. On the universality of inversion layer mobility in silicon MOSFET's: Part I-Effects of substrate impurity concentration, S. Takagi, A. Toriumi, M. Iwase and H. Tango, IEEE Transactions on Electron Devices, Volume 41, Issue 12, December 1994, pp. 2357–2362.
5. Physical Model of Incomplete Ionization for Silicon Device Simulation, Andreas Schenk, Pietro P. Altermatt and Bernhard Schmithúsen, 2006 International Conference on Simulation of Semiconductor Processes and Devices, September 2006, pp. 51–54.

Gate-All-Around (GAA) Nanowire n-FET Device Physics Based Performance Analysis

For gate-all-around (GAA) nanowire transistor, for improved gate to channel integrity and minimization of short channel effects (SCE), the nanowire diameter has to be reduced as amenable by the roadmap node up to 2–3 nm where later chapter will show nanowire will be transformed to stacked nanosheet architecture with gate wrapped up around. With reduced nanowire diameter, volume inversion is possible and as the doping of the nanowire is set in the vicinity of $10^{15}/cm^3$, the threshold voltage can be scaled as per voltage scaling so that there is enough inversion charge density with higher permittable gate overdrive or $(V_{gs}-V_T)$ where V_{gs} is applied gate voltage and V_T is threshold voltage of the GAA nanowire FET. The drain field penetration towards the source through channel surface and underneath the channel in a single gate architecture, is already eliminated by the gate-all-around architecture, specially with cylindrical gate rather than rectangular gate. High-K gate dielectric like HfO_2 is also an important part to enable high gate to channel integrity, volume inversion and uniform channel thickness when the consideration of drain voltage increase will be required for higher saturated drive current for this device. When this high-K gate dielectric is grown by versatile atomic layer deposition (ALD) method, the interface trap density D_{it} (traps/(cm^2–eV)) stays at a minimum level based on process sequence and ALD deposited high-K gate dielectric becomes the central process technology when the diameter of the nanowire goes below 5 nm as the D_{it} density keeps on increasing at the channel interface when the pore size is around 3 nm irrespective of lower nanowire substrate doping concentration and high-K gate dielectric. The second feature that becomes evident as the nanowire diameter is continuously reduced below 3 nm is quantum confinement. 2D quantum confinement, one way increases the energy level of lowest band (eV) and separates the other sub band in energy. This increases the threshold voltage of the device and decreases volume inversion through reduced gate overdrive. But

N. S. Ashraf, *Parameter-Centric Scaled FET Devices*, Synthesis Lectures on Emerging Engineering Technologies, https://doi.org/10.1007/978-3-031-84286-3_6

2D quantum confinement also decreases the transport related conductivity effective mass for the GAA structure, additionally, the splitting of sub-bands into higher energy levels, reduces the inter-subband-optical phonon scattering, a gain in transport mobility through reduced conductivity effective mass and reduced optical phonon scattering through sub-band. So, quantum confinement effects can ideally be used to gain on drive current of the GAA nanowire FET with a suitable reduced diameter setting. Both volume inversion through lower substrate non degenerate doping and quantum confinement reduce interface roughness scattering and boundary layer scattering, as the channel density is maximum at the nanowire pore center and also uniform across its boundary surface layer. Minimization of D_{it} through ALD grown high-k gate dielectric for GAA nanowire FET, also helps in reducing the off state leakage current which for higher gate integrity and lower substrate density with reduction of SCE, is already sufficiently reduced so that an I_{on}/I_{off} ratio of 10^5 can be achieved and the roadmap value of 10^6 can be targeted which will be better achieved by stacked nanosheet GAA FET to be discussed in the next Chap. 7. Minimization of D_{it} is also vital for improved subthreshold slope of gate-all-around nanowire FET, where use of high-K and GAA architecture systematically reduce the subthreshold slope equal to 60 mV/decade, the lower limit at T = 300 K in silicon or even in the vicinity of 30–40 mV/decade when the GAA FET architecture is converted to tunnel FET and negative capacitance (NC) Ferroelectric FET architecture, but, high process related D_{it} increase is noted in reference [1] for nanowire diameter close to 1 nm leading to aggravated subthreshold slope increase over 100 mV/decade [1, Table I].

Now we discuss the important reference [1] based on this book's author's viewpoints for key modeling requirements for GAA FETs as discussed above. As per [1] abstract's salient features, the authors of [1] confirms the requirement electrical conduction or channel density is concentrated near the center of the ultra narrow wire diameter. Another important observation from [1]'s abstract is nearly 200% increase of on current in silicon GAA nanowire FET for channel length L_g of 5 nm when compared to tri-gate silicon nanowire FinFET of 5 nm L_g, an I_{on} value of. 666 microampere/micron with the cited roadmap value of 900 microampere/micron [1]. P-FET devices generally has much lower on current density than n-FET due to their lower transport mobility and more susceptibility to optical phonon generation rate limited optical phonon scattering enhancement at higher gate vertical field and higher lateral field through drain when compared to n-FET. But, according to [1], GAA architecture with reduced nanowire diameter is equally suited to p-FET architecture through strain engineering, which affects both mean time between scattering and reduces the transport directed conductivity effective mass for hole in p-FET structure, leading to additional increase of drive current amounting to 973 microampere/micron [Table I, 1]. Underlapped gate to source junction, around 1 nm for $L_g = 5$ nm for GAA n-FET architecture, can increase the drive current I_{on} to 5854 microampere/micron and for GAA p-FET, 6125 microampere/micron [Table I, 1]. All these reported data are in ballistic mode quantum transport simulation. From cited reference articles, the authors in the paper [1] reveal that for thinner and lower dimension channel, lowers the permittivity of ε_{ch} of the channel through reduced screening of electron–electron interaction, but the impact of increased volume inversion for dimensional scaling is not discussed by [1] whether this effect plays a part in the screening. Another way of explanation of why scaling from 3 to 1D reduces the silicon permittivity manyfold because the 3D polarization charge volume or oriented dipole moments density are now significantly reduced in number density in 1D linear dimension leading to polarization loss and eventual reduction in silicon substrate dielectric permittivity. Due to dimensional scaling from 3 to 1D for silicon material, presence of strong quantum confinement effect shifts the lowest state band energy level in the conduction band up and sub-bands also increase in energy, possibly indicating a larger band gap due to quantum confinement in 1D nanowire diameter reduction and that may be another cause for reduced permittivity for 1D silicon nanowire material notwithstanding, some deviations are observed for III–V substrate based materials where a few larger band gap substrates tend to show near 3D dimension silicon permittivity of 11.7 ε_0 or slightly higher value deviating from this observation. So, the second corollary reasoning based on dipole moment density reduction in reduced volume of 1D silicon nanowire may be more proper intrinsic property based understanding for 1D silicon material's reduced permittivity value. In [1], quantum simulation based GAA nanowire performance characteristics as reported in Table I [1] is conducted by density functional theory (DFT) and non-equilibrium Green's function (NEGF) theorem. The author of this book, however found that in [1], the authors analyzed that the heavy doping of the channel significantly screens the electron–electron interaction, but the author of this book believes, the channel inversion concentration being high for that screening of electron–electron interactions, alternatively imply that the nanowire substrate doping must be lower as the both the requirements of thin channel dimension and volume inversion need to be met for this to happen which [1] did not clarify explicitly. In [1] also confirmed the author of this book's viewpoint that as nanowire diameter is reduced as per Moore's Law based node, the channel conduction film density shifts from the nanowire surface's boundary layer to innermost core, a much needed intrinsic device physical aspect of volume inversion with inversion charge density centroid right near the pore center and additionally causing the reduction of boundary layer scattering and surface roughness scattering. The source and drain doping density is determined to be $3 \times 10^{19}/cm^3$ but that is to reduce contact resistivity or making the contacts as ohmic as possible which [1] did not clarify fully. The contact regions should be abrupt, so that channel or substrate can be doped lower non-degenerate, otherwise as per [1], volume inversion may not happen even though the charge density centroid is moved to the core center, also uniform channel layer thickness is an issue with both gate bias increase and drain bias increase which [1] did not clarify that simply reduction of channel thickness due to nanowire pore dimension reduction, will not improve the transport efficiency of carriers in the channel for higher drive current. In [1] did not discuss the effect of increasing quantum confinement through nanowire diameter reduction on GAA FET's drive current performance which the author of this book discussed in the beginning paragraph sections prelude to [1] that quantum confinement can additionally boost the drive current which could have been studied by [1] by the simulation method the authors in [1] used.

In reference [2] further evidence of volume inversion and quantum confinement effect at room temperature corroborating the observation of the author of this book, has been reported by the authors in $In_{0.53}Ga_{0.47}As$ nanowire FET with nanowire dimension in 50 nm range. Due to lower transport mass and higher saturated drift velocity of in $In_{0.53}Ga_{0.47}As$ than silicon at T = 300 K, gate-all-around FET in $In_{0.53}Ga_{0.47}As$ can have very high mobility than scaled silicon GAA FET [2]. Therefore, when making inverter logic gate, the bottom stack of silicon GAA FET can be supplanted by $In_{0.53}Ga_{0.47}As$ GAA FET although lower D_{it} density by ALD grown high-K dielectric is a challenge for $In_{0.53}Ga_{0.47}As$ GAA FET [2]. There are other drive current instability and transconductance reduction factor in gate-all-around nanowire FET in silicon. Owing to the requirement of very low non degenerate doping density in the substrate, as L_g of nanowire is reduced below 5 nm and nanowire (NW) width accordingly, dimensional scaling to near 1 D and very thin channel layer results in very minute volume to determine the actual number of dopants in the channel. At room temperature or T = 300 K, considering 100% of all dopants, the actual number of dopants can be less than 100 when size dependent nanowire scaling is considered and this will lead to threshold voltage fluctuations through random dopant fluctuation (RDF) effect through dopant number and position in the depletion region. At T = 300 K, since the scaled V_T of the GAA FET is designed at a very low value by the gate-all-around structure, the fluctuations induced variance can be as much as 15%. When these GAA nanowire silicon FET are operated in ballistic mode in cryogenic temperature, severe incomplete ionization makes substantial percentage of the dopants in the substrate to be neutral. As a result, even though the dopant number is further reduced sizably from 100, the mean value does not change much for the threshold voltage V_T as this V_T is already a high value due to band gap increase of silicon at T = 4.2 K where these GAA silicon nanowire FET operates. But the due to very low number of dopants, the effect of their position and number variation on threshold voltage V_T variance can be more larger percentage at cryogenic temperature than room temperature. The concept of volume inversion, uniform channel thickness and the inversion channel charge centroid to be positioned near the center of the nanowire diameter, all break down into anomaly when this RDF effect is considered with size dependent nanowire L_g and width scaling, both at room temperature and at cryogenic temperature operation. Assigning negative bias at the source of the GAA nanowire FET, so that carriers are more injected from the source into the channel or making the substrate positively biased, can reduce the mean value of threshold voltage V_T but cannot substantially narrow down its variance and spread in room temperature and cryogenic temperature operation. So, channel layer inhomogeneity stays observable due to the RDF effect both at room temperature and cryogenic temperature operation. Under this circumstance when RDF effect cannot be made non-existent either by process or design modeling, size dependent quantum confinement effect can still provide the desired ballistic mode drive current through reduced transport mass and higher mobility through reduced inter sub band optical phonon scattering as stated before. At cryogenic temperature, use of non degenerate low substrate doping of the nanowire, results in large gate to substrate surface potential induced band bending and higher quantum confinement than room temperature. As a result, while higher inversion charge density and volume inversion effect along with uniform channel layer are affected by RDF, higher quantum confinement may additionally boost the drive current in the ballistic mode and $In_{0.53}Ga_{0.47}As$ GAA nanowire FET will have an advantage here on these transport related perspectives than silicon when n-FET logic devices fabrication are considered with higher drive current for giga scale microprocessor logic unit.

The other reliability feature of GAA FET architecture that reduces drive current and transconductance when room temperature operation is considered, is self-heating effect or SHE. With the ultra scaled nanowire length and width, silicon having a lower thermal conductivity, ballistic transport results in carriers travelling with their carrier temperature being elevated than room temperature and this causes degraded momentum with compounding self-heating effect where the carrier temperature dissipation does not happen normally through the channel when device is in saturation mode with high gate voltage and drain voltage bias. Quantum confinement effect elevates the inversion carriers to higher sub band in eV where the lowest band is also shifted up in eV and these high energy also results in carrier temperature increase leading to SHE based drive current decrease and transconductance decrease. For GAA nanowire FET, reduction of fields near the surface of nanowire circular or rectangular geometry, by reduced gate voltage and reduced drain voltage ensuring ballistic transport, are a solution for reduction of SHE effect or carrier temperature rise above than equilibrium substrate temperature. SHE effect is more dominant at cryogenic temperature operation, due to higher drain field penetration to the source through the equipotential neutral dopants that are not ionized and degree of ballistic mode transport is more intense along with carriers travelling with much higher kinetic energy and hence with much higher elevation in their carrier temperature which has to be decreased over a steep slope of T = 4.2 K substrate temperature than normal T = 300 K substrate temperature. The next Chap. 7 where gate all around nanowire now turns into stacked nanosheet, these reliability effects will be further discussed through relevant reference articles along with possible technology based remedies, although there will be RDF and SHE effects in the size dependent NW scaled GAA FETs at possible 1–2 nm dimension.

References

1. Performance Limit of Gate-All-Around Si Nanowire Field-Effect Transistors: An *Ab Initio* Quantum Transport Simulation, Shiqi Liu, Quihui Li, Chen Yang, Jie Yang, Lin Xu, Linqiang Xu, Jiachen Ma, Ying Li, Shibo Fang, Baochun Wu, Jichao Dong, Jinbo Yang and Jing Lu, Physical Review Applied, 18, 054089, 2022, pp. 054089-1–054089-15.
2. Size-Dependent-Transport Study of $In_{0.53}Ga_{0.47}As$ Gate-All-Around Nanowire MOSFETs: Impact of Quantum Confinement and Volume Inversion, Jiangjiang J. Gu, Heng Wu, Yiqun Liu, Adam T. Neal, Roy G. Gordon and Peide D. Ye, IEEE Electron Device Letters, Volume 33, Issue 7, July 2012, pp. 967–969.

Gate-all-around (GAA) nanowire FET architecture has been analyzed with device physics based transport assessment. In this chapter, the author of this book shows the configuration of vertically stacked gate-all-around nanowire nanosheets where the term 'nanosheet' arises from FinFET based heigh reduction while the nanowire size (width and length) is controlled when the node is around 5 nm. Since superior gate to channel integrity is preserved in vertically stacked nanowire gate-all-around nanosheet FET architecture also, the author first takes the reference from classic article [1] "Design Insights of Nanosheet FET and CMOS Circuit Applications at 5-nm Technology Node" by the authors. As the authors stated in [1] that as the feature length is aggressively scaled as per Moore's Law beyond 5 nm node, the GAA FETs with vertically stacked nanosheet allow a multiple number of channels to be used amenable to volume inversion with their chosen substrate density $10^{15}/cm^3$ for superior drive current performance for logic circuits applications which is the main focus of this book from the FET applications perspective. As the authors in [1] rightly observe that GAA nanosheet with perimeter enhanced effective gate width W_{eff} also plays a role in channel homogeneity when the usual quantum confinement effect makes the channel layer close to the nanowire surface, inhomogeneous and also adds in overall drive current enhancement for thin sheet like fin height and ultra small node nanowire diameter or L_g. For different logic core circuits, as per the requirement of drive current and leakage current, this nanosheet (NS) width adjustments also allows CMOS compatible circuit layout to be constructed for higher power driving logic processing units [1]. Whereas for GAA nanosheet FET in [1], the authors used gate length L_g to be between 5 to 16 nm, sheet thickness or fin height 5 to 10 nm, they widened the fin width or sheet width between 10–50 nm as vertical stacking will allow this range to be accommodated with area wise minimal increment of device lay-out. Length of underlapped spacer to gate to source junction is kept at 5 nm in [1] which in Chap. 6, we found through [2] that the drive current/micrometer increases by quite a proportion in the ballistic mode. For this stacked GAA nanosheet (NS) FET in [1], self-consistent Poisson's and Schrödinger's equations as defined by non-equilibrium Green's Function (NEGF) have been used for various performance features assessments. In [1] for simulation purposes, the nanowire length or node 5 nm was converted to an equivalent linear length of 16 nm and nanosheet width of 10 nm was converted to an equivalent perimeter controlled width of 60 nm for an overall threshold voltage at T = 300 K to be 330 mV. The source and drain doping for near ohmic type contact purposes were kept at 2×10^{19}/cm^3 where in line with the observation of the author of this book, the authors in [1] selected the substrate doping of nanosheet to be 10^{15}/cm^3. The height of the metal gate was chosen to be 60 nm although with this large height metal gate granularity (MGG) effect may change the work function of metal which may affect the flat band shift of the nanosheet and shift and variation in their mentioned value of threshold voltage which the author of this book finds an important parameter to be included in the simulation results in [1] which was not the case. As these stacked nanosheet requires buried oxide (BOX), a 50 nm BOX is used while the high-k gate dielectric was HfO$_2$ with effective oxide thickness (EOT) of 0.78 nm. For quantum confinement effect simulation, the authors in [1] employed density-gradient drift diffusion equations where the inhomogeneous distribution of inversion electrons in GAA nanosheet n-FET near interfacial SiO$_2$-HfO$_2$ high K gate dielectric and nanowire interface, can be properly accounted for as in [2], we saw that increasing quantum confinement increases the volume inversion effect by pushing the channel density centroid towards the nanowire pore center with which different nanosheet in this case are constructed. As reported by authors in [1], the saturation bias for this nanosheet FET are $V_{gs} = 0.7$ V and $V_{ds} = 0.7$ V whereas for transfer curve analysis, the drive current was taken as $V_{gs} = 0.7$ V with V_{ds} being 0.04 V. The authors in [1] for I_{ds}–V_{ds} saturation bias curve point, extracted a value of drive current to be 62.1 µA which as per [2] for 10 nm nanowire width will be 62.1 µA/0.01 µm = 6210 µA/µm meeting or exceeding the roadmap value of 900 µA/µm mentioned in [2] for GAA nanowire FET only. As further reported in [1], the subthreshold slope (SS) for this n-channel nanosheet GAA nanowire FET was 62 mV/decade, although the process related formation of interface trap density D_{it} from fe 10^{11}/cm^2–eV to 10^{12}/cm^2–eV, may increase this value of SS and therefore both the underlying thin interfacial SiO$_2$ with HfO$_2$ high-K gate stack as grown with atomic layer deposition (ALD) must be used and the author of this book assumes that the authors in [1] verified from the experimentally calibrated data supplier that ALD was used in high-k HfO$_2$-interfacial SiO$_2$ deposition process. As reported in [1], the nanosheet n-channel FET architecture has lower DIBL of value 30.69 mV/decade which may rise a bit when ballistic mode current is determined where the drain voltage has to be lifted so that n-FET is in saturation with observation by the author of this book. In [1], for linear I_{ds}–V_{gs} transfer curve bias, n-channel nanosheet FET provided a drain current of 62.2 µA which means ballistic mode current was prevailing at a lower drain voltage of

0.04 V from 0.4 V for the saturation current. This increase in nanosheet drive current is due to its larger effective sheet width $_{eff}$ providing better spatial current averaging and more dense channel formation as per volume inversion effect. In [1], the subthreshold leakage current I_{off} was reported to be for n-channel nanosheet GAA FET as 1.89 pA, so the I_{on}/I_{off} ratio is = 62.1 μA/1.89 pA, almost 33×10^6, highly promising for roadmap at node 5 nm where in the vicinity of I_{on}/I_{off} ratio 10^6 or better is generally prescribed. Vertical stacking of GAA nanosheet FET architecture enhances the quantum confinement effect and hence leakage current is further reduced. The authors further showed that fin height or nanosheet thickness (NS_H) has to be thinned towards low end or 5 nm for better performance of all parameters so far described from [1]. The authors also found in [1] that lower nanosheet width NS_W is beneficial to control I_{off}, subthreshold slope SS and DIBL to a lower value. In [1], the authors showed that I_{on}/I_{off} ratio degrades at higher temperatures than 300 K but cryogenic nanosheet FET performance where ballistic mode is more prevalent for near intrinsic bulk doping of the order of $10^{15}/cm^3$ so that superior drive current and I_{on}/I_{off} ratio is expected, probably demanded comprehensive simulations at temperatures as low as 50 K and where various incomplete ionization related impacts on carrier density through threshold voltage and transport mass change or reduction and scattering time enhancement need to be assessed in presence of neutral impurity scattering, obviously behind the scope of the article [1] where high temperature transport features as reported were not as complex and intrigued as low temperature or cryogenic temperature operation of nanosheet n channel FET.

In [3], the authored showed through Verilog-A compact modeling that for mobility enhancement of nanosheet GAA p-FET, the requirements are low transverse field or gate voltage but enough inversion charge density necessitating lower substrate doping concentrations in the $10^{15}/cm^3$ as with higher transverse field and lateral field, with higher inversion channel density, the mobility degrades due to higher transverse field and reduced sheet thickness (fin height) induced quantum confinement effect which increases the transport conductivity effective mass of hole, causes additional optical phonon scattering and interface roughness scattering with increasing gate overdrive induced increment in channel carrier sheet density and thus decreases mobility [3]. In [3], the author of this book holds the opinion that non equilibrium mixed-mode electro thermal simulation if coupled with Verilog-A simulation program, might have revealed whether the carrier temperature has risen with respect to substrate temperature when gate overdrive is high and channel sheet density is also high. Rise in carrier temperature with thin nanosheet thickness (fin height) as a result of quantum confinement, may cause the 2D transport effective mass from (100) direction transport and (110) sidewall directed transport to increase and this explanation was lacking in [3]. Vertical nanosheet stacking also causes multiple threshold voltages in nanosheet n-channel FET, inducing varied degree of screening of ionized dopants by the channel inversion sheet charge density and that may affect mobility variation, if due to the thinness of nanosheet and vertical nanosheet stacking number, the channel sheet density is high but their layer thickness is reduced, causing reduced screening of ionized dopants and therefore, decrease in mobility. In [3] as the authors mention that due to thinness of nanosheet (fin height), induced quantum confinement effect increases the transport directed conductivity effective mass, it means in the conduction band due to valley splitting and degeneracies for increasing quantum confinement, larger transport conductivity effective mass in the transport direction would mean that the lowest energy band in eV is closer to conduction band minima E_c and the split higher energy level sub bands are closer to each other in eV and hence, with increasing gate overdrive as the sheet density in the channel increases [3], not only the lowest band is occupied by electrons but also the higher sub bands are also occupied by some electron concentration and this generates inter sub band optical phonon scattering enhancement which in turn reduces the channel mobility. If on the other hand, other than silicon, III–V material is used, quantum confinement and decreased nanosheet thickness, in some cases reduces the transport directed conductivity effective mass. So now the lowest band is further shifted from the conduction band edge E_c and other valley-split sub band with degeneracies are shifted further apart higher in energy levels. In this case, the most electron concentration of sheet channel density lies in the lowest band and higher separation of energy levels of the sub bands mean that inter sub band optical phonon scattering is reduced and this will mean less proportionate decrease of channel mobility with increasing sheet density reported in [3]. The fact that although nanowire GAA FET and nanosheet GAA FET both have GAA architecture for better gate to channel integrity and shorter effective 2D scaling length at moderate gate and drain bias, nanowires are circular and hence amenable to complete volume inversion for lower substrate doping concentration, but nanosheet are characterized by sheet width and sheet thickness, where slightly wider sheet gives higher drive current and also drift mobility [3] but for nanosheet, the required thinness nanosheet or fin height in a FinFET like structure, does not allow complete volume inversion even though the inversion charge density like nanowire core, is shifted from the surface in nanosheet GAA FET, the inversion layer thickness is still non uniform with inhomogeneous distribution from the nanosheet high-k gate dielectric-silicon surface towards the depth of sheet thickness. Taller fin like structure or when the nanosheet thickness is relaxed or enhanced, quantum confinement effect is reduced and the channel carriers face increasing scattering at the surface of nanosheet when both vertical field and lateral field are high [3] and also the thin nanosheet in fully depleted body or SOI structure may be violated due to the finite depletion width due to the substrate doping, causing extra series resistance from the depletion region bottom to the substrate contact but thinner nanosheet will always have the silicon body fully depleted with low enough body or substrate doping concentration. So as per [3], at low transverse field, adjusting body doping, nanosheet width and nanosheet thickness, ballistic mode transport may still be achieved at T = 300 K with moderate drain voltage even though sheet channel inversion density is lower. Other geometry effect that can be tried through process is contouring circular type arcs at the corners of four edges of nanosheet, that will increase volume density or uniformity of channel thickness even though nanosheet is thinned or fin height is reduced. In [4], the author of the referred article conducts a comparison between different reliability effects like negative bias temperature instability (NBTI) for nanosheet p-FET, positive bias temperature instability (PBTI) for nanosheet n-FET, hot carrier degradation (HCD), time dependent dielectric breakdown (TDDB) and self heating effect (SHE) for both n and p nanosheet FET with channel carrier transport direction mainly through (100) silicon surface orientation of nanosheet top and bottom with additional conduction taking place through (110) oriented sidewalls and arcuated corners. Reliability effects were focused in [4] for sheets having arcuated corners as suggested by the author of this book, in relation flat top bottom nanosheet configuration and [4] showed that arcuated corners do not always result in better FET performance and some reliability features are degraded comparably when conduction mode is in (100) direction compared to (110) direction with sheet having arcuated corners and number of vertically stacked nanosheet also play a role in overall reliability assessment [4] where the SHE effect is more conspicuous as the carrier temperature rise does not have a relaxation path through the substrate contact and SHE effect is more pronounced for decreased sheet thickness [4] and increasing surface vertical and lateral field where ballistic mode transport is desired. Contrasting with [1, 4] reported that wider nanosheet results in more SHE effect aggrandizement. Although, the author of this book believes that as per [1], wider nanosheet results in uniform channel thickness and deeper channel thickness ameliorating the gate vertical field and drain lateral field induced carrier temperature enhancement than equilibrium substrate or lattice temperature as observed in SHE. In [4] also reported that for n-channel nanosheet GAA FET, the electron mobility is higher than peak mobility in (100) direction compared to (110) direction whereas, in p-channel nanosheet GAA FET, hole mobility in (100) direction is lower than (110) direction with its value attenuated from hole peak mobility in silicon. As per [4], the dangling bond density Si:H is higher in (110) surface than (100) surface and that means that interface trap density D_{it} will be higher in (110) sidewall conduction in nanosheet FET than in (100) direction transport which may cause instability in threshold voltage variation and its degradation most for bias temperature stress (PBTI and NBTI). In [4] through simulation, the author showed importance of controlling the radius of arcuated corners of the nanosheet edges and sidewall profile optimization for reliability improvement. It may be inferred here that the effect of line edge roughness (LER) is directly related to sidewall profile optimization which impact the critical dimension (CD) variation in across die and from die-to-die causing a six sigma threshold voltage variation profile and this effect is not reported in [4]. Self heating effect (SHE) mentioned in [4] for T = 300 K, gets accentuated in cryogenic temperature or around T = 4.2 K as has been discussed in previous Chap. 6 for GAA nanowire FET architecture, as with higher ballistic mode transport at cryogenic temperature, significant carrier heating takes place and the slope of temperature decrease is higher as the substrate temperature is in cryogenic cooling region or T = 4.2 K. 2D scaling length optimization to make it as small as possible, lower vertical gate field, lower lateral drain field and uniform and higher channel thickness rather than higher channel sheet density, can be combined to tackle the effect of SHE on drive current reduction for logic circuits and transconductance degradation for AC circuits. Another important article [5] that aims to mitigate self-heating effect (SHE) through process optimized modification of isolation buried oxide (BOX) by buried oxide accompanied by a crystalline-diamond-like carbon (DLC) substrate material (possessing higher thermal conductivity than buried oxide) is intimately placed along underneath the lower nanosheet layer. This results in a heat dissipation path and reduces the carrier temperature from the channel in the nanosheet to the DLC substrate. SHE can be also reduced by a lower transport conductivity effective mass nanosheet GAA FET material, where with increasing quantum confinement due to thinness of nanosheet, if the transport directed conductivity effective mass further lowers, only the lowest band in the conduction band is occupied which is shifted up in energy from the conduction band edge E_c eliminating partially surface roughness scattering and boundary layer scattering and then if the sub-bands are also shifted further up in energy levels due to this decreased conductivity effective mass of the material with which nanosheet n-channel GAA FET is fabricated, reduced inter sub band optical phonon scattering will mean that all the three commonly reported scattering events causing carrier temperature rise in the channel during saturation mode conduction are subdued and as a result, carrier temperature increase will be less or gradual than substrate temperature leading to reduction of SHE. Next in [6], a core insulator SiO_2 embedded nanosheet structure has the potential of shifting the larger part of the channel density away from the surface to the middle centered core layer SiO_2-nanosheet interface, suggesting nanowire like volume inversion and reduction of surface roughness scattering and boundary layer scattering and accordingly reduction of self-heating effect (SHE). [6] shows the way of configuring an ultrathin 2 nm SiO_2 core embedded between the nanosheet thickness or height. This is similar to ultra-thin-body (UTB) SOI with a thin buried oxide (BOX) but what has not been reported in UTBSOI is that after the edge of bottom of BOX, there is a small silicon thickness and then the bottom contact is formed which we call back gate that can control the short channel effect and 2D scaling length extension from the drain through the thin body to the source by suitable control of the back gate (in accumulation, inversion etc.). Basically in [6], the original nanosheet thickness or fin height is further reduced into two partial parts by the core insulator SiO_2 embedded structure. Both the top nanosheet and bottom nanosheet gate electrode vertical field terminate on the insulator top edge for top gate and insulator bottom edge for bottom gate. Along with quantum confinement enhancement for partial splitting of the nanosheet thickness in [6], volume inversion near the edges of core embedded SiO_2 (top and bottom) is verified through Silvaco ATLAS TCAD simulation by the authors of the paper [6] and lower subthreshold leakage current but similar saturation bias drive current are reported in [6] when compared to nanosheet FET only for same nanosheet length $L_g = 12$ nm but with sheet thickness of 5 nm when core insulator embedded vertically stacked nanosheet FET structure is used instead of vertically stacked nanosheet FET structure only. The drive current in microampere value is also larger in value than [2] when I_D–V_{gs} is measured and simulated in [6] but [6] has a nanosheet width of 20 nm and larger nanosheet length, so this structure needs to be further scaled for 2–3 nm node adjustment with sheet thickness 3 to 5 nm. In [6] core insulator can be also high-K dielectric, with additional benefit of reducing leakage current but as per [6], the interface trap density near high-k dielectric and nanosheet interface D_{it} can rise as much as $10^{12}/cm^2$–eV degrading subthreshold slope value from simulated value near 65 mV/decade [6]. It can be also inferred from device physical analysis that the author of this book explores on findings in [6] that due to the core embedded thin 2 nm insulator SiO_2 nanosheet FET, when the top gate electrode is positively biased, in the nanosheet substrate which has been doped with acceptor dopants, the mobile holes are pushed down but they cannot terminate on the substrate contact in usual nanosheet configuration, rather these holes accumulate on the core embedded thin insulator top interface. In the process through the insulator, this sheet of holes induces negative electron sheet charge density on the bottom edge of thin core embedded insulator. Now when volume inversion occurs in the nanosheet, the large electron density of the channel that are at the interface of top insulator-nanosheet interface, recombine with the excess mobile holes making the top core embedded insulator edge almost equipotential with only having electron channel concentration filled with from top of the nanosheet surface to the top of the core embedded thin insulator surface. When the bottom gate is positively biased, this time mobile holes of the acceptor dopant of the nanosheet, are pushed towards the bottom edge of core embedded insulator. Through the insulator these holes then induce a sheet of negative charge density on the top edge of the core embedded insulator. The quantum confinement effect places enough inversion channel carriers near the bottom edge of the core embedded thin insulator at the center of nanosheet thickness. As a result of which, through recombination this extra hole density is extinguished by the inversion channel electron density. So, now the bottom edge of core embedded insulator is also under heavy electron concentration channel density and forms almost equipotential surface. In the design process of [6], it needs to be ensured that the vertical field through the gate from top electrode does not penetrate through the top nanosheet thickness and tunnel through the ultrathin core embedded oxide top edge to the bottom edge, eventually degrading the 2D scaling length and same is true for the bottom electrode, where the vertical gate field now penetrates through the bottom nanosheet thickness by extra tunneling from the bottom edge of the core embedded insulator to the top edge of the core embedded ultrathin insulator. This will also exacerbate the 2D scaling length.

For cryogenic vertically stacked GAA nanosheet structure, with superior drain voltage and gate voltage, ballistic mode conduction can still result in high drive current even though for low non degenerate body doping of the nanosheet, considerable incomplete ionization result, which causes discreetness of channel thickness from the sheet surface towards the sheet core or center. The maximum gain in mobility and saturated drift velocity for nanosheet GAA FET operating at cryogenic temperature comes from (1) reduced phonon scattering, (2) reduced optical phonon scattering although the discreteness of the channel may cause surface roughness scattering and boundary layer scattering to be enhanced at reduced nanosheet thickness, raising the carrier temperature substantially above than substrate temperature near T = 4.2 K. This will result in self-heating effect and previous discussion focused on how to combat reduction of SHE from device structural and substrate material perspectives including the method shown in [5] which eliminates SHE at T = 300 K and should minimize it for T = 4.2 K.

References

1. Design Insights of Nanosheet FET and CMOS Circuit Applications at 5-nm Technology Node, V. Bharath Srinivasulu and Vadthiya Narendar, IEEE Transactions on Electron Devices, Vol. 69, No. 8, August 2022, pp. 4115–4122.
2. Performance Limit of Gate-All-Around Si Nanowire Field-Effect Transistors: An *Ab Initio* Quantum Transport Simulation, Shiqi Liu, Quihui Li, Chen Yang, Jie Yang, Lin Xu, Linqiang Xu, Jiachen Ma, Ying Li, Shibo Fang, Baochun Wu, Jichao Dong, Jinbo Yang and Jing Lu, Physical Review Applied, 18, 054089, 2022, pp. 054089-1–054089-15.
3. Compact Model for Geometry Dependent Mobility in Nanosheet FETs, Avirup Dasgupta, Shivendra Singh Parihar, Harshit Agarwal, Pragya Kushwaha, Yogesh Singh Chauhan and Chenmiing Hu, IEEE Electron Device Letters, Vol. 41, Issue 3, March 2020, pp. 313–316.
4. A Review of Reliability in Gate-All-Around Nanosheet Devices, Miaomiao Wang, Micromachines, 2024, 15, 269, pp. 1–20.
5. Demonstration of a Nanosheet FET With High Thermal Conductivity Material as Buried Oxide: Mitigation of Self-Heating Effect, Sunil Rathore, Rajeewa Kumar Jaisawal, P. N. Kondekar and Navjeet Bagga, IEEE Transactions on Electron Devices, Volume 70, Issue 4, April 2023, pp. 1970–1976.
6. Core-insulator embedded nanosheet field-effect transistor for suppressing device-to-device variations, Donghwi Son, Hyunwoo Lee, Hyunsoo Kim, Jae-Hyuk Ahn and Sungho Kim, Scientific Reports, 14, Article Number: 7462 (2024), pp. 1–7.

Conclusion and Future Remarks

8

Most of the modeling analysis of nanoscale various FET architectures for n-channel and p-channel FET that are configured on advanced channel length nodes below 7 nm, do not take into account precise determinations of various FET and material centric transport parameters that have been elucidated and computed in this Book by the author in the Chaps. 1–5. First of all, for non degenerate doping, from the published articles in last many years, it is the general understanding that the true ionization percentage of the activated acceptor dopants in n-channel FET and activated donor dopants in p-channel FET, there lies a conceptual misconception that for moderate non degenerate doping in the range 10^{15}–10^{16}/cm^3, the actual ionized dopants are always 100% but in this Book, for the first time through Chaps. 1 and 2, it was illustrated that beyond 10^{17}/cm^3 donor and acceptor doping in silicon at T = 300 K, the ionization gradually goes down in line with reference articles and textbooks but after reaching a minima around few 10^{18}/cm^3 for both acceptor and donor doping silicon, the activated dopant ionization percentage goes up and for fully degenerate doping in the range 10^{19}–10^{21}/cm^3, the ionization is 100% where the analytical formulas in textbooks and reference articles (the author of this book is excluding the citation index of these articles and textbooks as this conception is a conventionally yet erroneously accepted) actually point to very low ionization percentage at degenerate doping for both donor and acceptor doping in silicon at T = 300 K. Moreover, the activated ionization of acceptors like boron in silicon is lower than phosphorous in silicon at T = 300 K between non degenerate doping values 10^{17}–10^{19}/cm^3, although for degenerate doping of the order of 10^{20}–10^{21}/cm^3, both for boron doped acceptor silicon material and phosphorous doped donor silicon material almost so no difference in ionization values pinned at 100% values at T = 300 K. Chapter 2 of this book extends these calculations using degeneracy induced non parabolicity in the valence band maxima for p-type silicon

N. S. Ashraf, *Parameter-Centric Scaled FET Devices*, Synthesis Lectures on Emerging Engineering Technologies, https://doi.org/10.1007/978-3-031-84286-3_8

and conduction band minima for n-type silicon at T = 300 K. Non parabolicity for both n-type and p-type silicon material, increase the effective density of states in the conduction band and valence band by increasing the 3 D density of states of effective mass for electron in n-type silicon and hole in p-type silicon in a non linear manner taking the effect of increasing non parabolicity for highly degenerate doping crossing 10^{19}/cm^3 and reaching up to 10^{21}/cm^3. Proper knowledge of the activated dopant in n-FET or p-FET at a chosen substrate doping density is important for FET bulk potential and threshold voltage calculation and eventually gate over drive related inversion channel density calculation for different drain voltage bias at T = 300 K. This inversion channel density whose thickness spatially varies from the source end to the drain end of the FET, with thickness wider at the source and narrower the at the channel pinch-off point, which is considered for saturated drain current calculation, although the pinch off point rapidly vanquishes in the time scale of carrier transport as the highly constricted stream of electrons drift with high saturated drift velocity in n-channel FET attracted by the drain positive potential with quick widening of the layer near the drain just like spill-over which is also detected at ultra short time scale high speed and high field transport. But, the point is that if the ionization differs from 100% for the chosen substrate doping silicon even near 10^{17}/cm^3 for n-FET and p-FET for T = 300 K, this will need to be properly analytically computed and transformed to threshold voltage calculation, so that precision carrier density for vertical and lateral field can be determined and the eventual drain current from subthreshold to threshold to saturation bias. The author did not see previous papers composed in the 1970s and 1980s and 1990s and 2000s when conventional MOSFET architecture was the norm, that precise ionization percentage was calculated from low end of non degenerate doping to high end of non degenerate doping (in silicon substrate at T = 300 K) to combat short channel effects with the precise analytical equations based approach narrated in this Book in Chaps. 1 and 2. For advanced FET architectures, TCADs like Silvaco and Synopsys ATK and DFT analysis along with full 3 D quantum simulation methods are employed for drain current calculation, subthreshold current calculation, I_{on}/I_{off} ratio, channel mobility and saturated drift velocity at T = 300 K. The author of this book is assuming that precise band structure based transport parameters are extracted for gate voltage bias and drain voltage bias, for instance effective density of states of conduction and valence band, 3D density of states effective masses of electron and hole, conductivity effective masses of electron and hole that determine channel mobility and saturated drift velocity, different scattering events like (1) ionized impurity scattering, (2) thermally phonon limited scattering, (3) optical and non optical phonon related scattering and (4) interface or surface roughness scattering. Full band DFT and 3 D full band quantum simulations are able to compute these scattering times as a function of carrier drift energy that increase their momentum or kinetic energy from lateral field increase through the drain bias. Yet for silicon at T = 300 K, Silvaco and Synopsys Quantum ATK TCADs report erroneous values for 3D density of states effective masses for electron and hole in silicon and also for conductivity effective masses for electron and hole in silicon (Again, the author of this book is excluding the reference indexes of these extracted values reported by TCADs as demonstration tutorials). DFT based approaches and 3D full band quantum simulation based works on advanced silicon FETs have not explicitly reported the 3D density of states effective mass of electron as 1.18 m_0 and hole as 0.81 m_0 for silicon at T = 300 K [1, 2] (again, the author of this book is excluding the citation indexes of these articles as the focus in these articles were on scaling based drain current improvements and other Roadmap based benchmark parameters rather than systematically quoting the 3D density of states effective masses and conductivity effective masses at T = 300 K used in their simulation methods). The author of this Book used these two computed effective mass values at T = 300 K from T = 4.2 K, from a temperature dependent effective mass calculation and also calculated at T = 4 K and T = 300 K as referred in Tabular values in [1, 2], in Chap. 1 till 5 of this Book for all FET related device parameters calculation that ultimately define FET drain current in the saturation mode.

Transport related important parameters of n-FET and p-FET like inversion channel mobility has been calculated as a function of vertical gate to channel field for a particular doping ranging non degenerate to degenerate doping and also inversion channel mobility plotted as a function of substrate doping from non degenerate level to very high degenerate doping level. Imprecise determination of conductivity effective mass for electron in n-FET and hole for p-FET along with variation of determination of overall scattering time by a combination of various scattering events can result in erroneous mobility values for saturated drift current calculations for n-FET and p-FET. Chapters 3 and 4 in this Book by the author shows how to calculate transport related conductivity effective masses for electron and hole as a function of substrate doping from non degenerate to degenerate levels taking account of degenerate doping induced band non parabolicity. The overall scattering time is also extracted in Chap. 3 from majority carrier mobility data as a function of substrate doping up to degenerate level and additionally, for minority carrier mobility data as a function of substrate doping up to degenerate level taking from the reference discussed in Chap. 3 of this book. Now, how to determine whether an inversion channel mobility reported as peak value owing to temperature limited phonon scattering in silicon at T = 300 K for a particular substrate doping value? First, we need to locate the mobility peak value from inversion channel mobility as a function of vertical gate to channel field point for a particular substrate doping. Then using the equation derived in Chap. 3 of this book for conductivity effective mass for electron in n-FET as a function of majority carrier substrate doping, we can know the overall scattering time from the Drude mobility equation. Since, at mobility peak at a particular gate to channel field, the scattering event is temperature limited phonon mobility, then this scattering time gets calculated by this Drude equation with the computed effective conductivity mass for electron. Now through DFT and full band 3D quantum simulation along with more accurate Ensemble Monte Carlo simulation, this scattering time can be recalculated at that particular gate to channel vertical field point. If there is discrepancy, then we know that DFT based and full band 3D quantum simulation or possibly Ensemble Monte Carlo simulation are not giving both proper scattering time value and proper conductivity effective mass value for electron so that inversion channel mobility from simulation at that particular vertical field can be taken as accurate. Already from Chap. 3 conductivity effective mass calculations for electron and hole in silicon when these carriers are majority and minority carriers for silicon material at T = 300 K expose in the inaccurate determinations in TCAD software like Silvaco and Synopsys quantum ATK. This anomaly will be clear to the readers of this book as they go along systematically derived parameters important for majority carrier FET and minority carrier FET by derived analytical equation based values. DFT, full band 3D quantum simulation can be more trusted for high field transport or ultra short scale devices with high lateral field, where the 3D density of states effective masses of electron and hole and conductivity effective masses of electron and hole become function of not only degenerate doping through band non parabolicity effect but also by high momentum carrying lifted values of carrier kinetic energy in the conduction band for n-FET and valence band for p-FET. Quantum confinement induced changes in conductivity effective mass in 2D confinement direction will need to be recalculated through band structure effects from 3D conductivity effective mass and an analytical derivation of this 2D conductivity effective mass as a result of quantum confinement is difficult to be precisely defined unless DFT generated and full band 3D quantum simulation generated 3D conductivity effective mass values of electron and hole are not as precise as reported in Chap. 3 of this book, particularly, for nanosheet and gate all around nanowire FET where non degenerate substrate doping is used and quantum confinement effect is important for drain current determination in both these FET architectures at advanced gate nodes.

In the published articles that report inversion channel mobility for n-FET and p-FET considering conventional FET architectures, most channel mobility versus substrate doping focuses on ionized impurity scattering with a level of decreased screening of ionized dopants in the depletion region as the inversion channel thickness gets smaller when the doping is increased from low non degenerate to moderate non degenerate to degenerate doping concentrations. Therefore, in these reported plots of the article it needs to be determined the maximum gate voltage overdrive ($V_g–V_T$) where V_T, the threshold voltage of FET keeps on increasing for increasing substrate doping density, hence the overdrive factor decreases for increased substrate doping with an indication that inversion carriers induced screening of depletion region dopants is reduced, causing the natural decrease in mobility first gradually and then with a larger slope for very high substrate doping concentrations. Beyond this maximum gate voltage overdrive allowed as the doping in the substrate is increased, the natural ionized impurity scattering will lead to optical and non optical phonon scattering and then interface roughness scattering and at extreme degeneracy, carrier-carrier scattering as the inversion channel thickness becomes continuously entrenched for increasing substrate doping concentration with the general assumption of low drain voltage, so that lateral field effect on the drift mobility is negligible. It is the author of this book's opinion that most of the reported channel mobility-doping concentration plots of articles for aggressively scaled n-FET and p-FET do not provide this analytical information in the content of the articles' write-ups. Also, due to doping, the conductivity effective mass of electron and hole for carrier transport in n-FET and p-FET change and generally increase due to the band non parabolicity factor with increase of substrate doping in these FETs. Silvaco, Synopsys quantum ATK, DFT and full band 3D quantum simulation all should be therefore precisely calculating these conductivity effective masses of electron and hole for inversion carrier transport in n-FET and p-FET when plotting inversion channel mobility as a function of substrate doping considering ionized impurity scattering. DFT, full band 3D quantum simulation and Ensemble Monte Carlo simulations all should be able to calculate the scattering time due to ionized impurity scattering as a function of carrier of energy translated from drift momentum from low lateral field transport perspective. Then the Drude mobility equation can be used to calculate the inversion channel mobility in analytical form from conductivity effective mass data and overall scattering data where the gate over drive level will indicate whether the overall scattering data is exclusively inferring the ionized impurity scattering. This Drude mobility equation then can be checked with simulation extracted channel mobility data at a particular substrate doping point using the above transport based simulation methods mentioned. Chapters 3 and 4 of this book by the author, show the analytical equations that compute majority carrier conductivity effective mass of electron and hole as a function of substrate doping from non degenerate to degenerate including band non parabolicity, minority carrier conductivity effective mass of electron and hole as a function of substrate doping from non degenerate to degenerate including band non parabolicity, majority carrier scattering time of electron and hole and minority carrier scattering time of electron and hole from non degenerate to degenerate including band non parabolicity and majority and minority carrier mobility of electron and hole as a function of substrate doping from non degenerate to degenerate including band non parabolicity all for silicon material at T = 300 K.

A very precise and definitive way to calculate ionized dopant percentage in any material including wholesomely researched silicon material at T = 300 K both for n and p-type dopants, is to first experimentally extract (1) Hall mobility and (2) Hall conductivity. Then the ionized free carriers concentration can be extracted from the equation, $\sigma_H = qn\mu_H$ for n-type material where σ_H is Hall conductivity, μ_H is Hall mobility based on ionized dopants only and n is the ionized free carrier in a n-type substrate material being silicon or any other material. The equation for p-type material can be analogously developed with n parameter substituted by p parameter. Knowing the actual doping concentration of the substrate N_D or N_A, now the ionization ratio n/N_D can be determined for any material at T = 300 K from non degenerate doping value to high degenerate doping values. Same can be done for p/N_A at T = 300 K for any material from non degenerate doping value to high degenerate doping values. This ionization percentage determination is pivotally critical from device and material modeling perspective for all FET based architectures, since silicon does show a variation with incompletely ionized dopants between $10^{17}/cm^3$ and $10^{19}/cm^3$ with almost full ionization from some $10^{19}/cm^3$ to $10^{21}/cm^3$ and this difference of ionization percentage plays a role in inversion channel density in a FET for a particular gate and drain bias, threshold voltage variation, inversion channel mobility variation from ionized impurity scattering as well as neutral impurity scattering, saturated drift velocity, subthreshold leakage current, saturation drive current, etc. Unfortunately, to-date we do not see this experimental method being successfully applied to other materials like group IV materials, Ge, Diamond and Carbon, many of the high speed materials III–V based substrates, 2D materials, carbon nanotube, graphene, hexagonal boron nitride, etc. Because based on the band structures of these materials for n-type and p-type dopants in these materials at T = 300 K, the activation energies of these dopants relative conduction band and valence band edge, band gap of these materials and how the increased degenerate doping concentration change the discrete activation energies in n-type and p-type substrate materials into continuum band forming overlapping with conduction band or valence band that may determine how close is the activation percentage of dopants near 100% as we observed for silicon at T = 300 K for very high degenerate doping concentration of the substrate. The authors who have researched on these alternative materials than silicon heavily depended upon DFT and full band 3D quantum simulation where this free carrier density at a particular substrate doping is intrinsically determined in the course of simulation flow where all the relevant parameters taking the precise 3D band structure into account are computed and integrated with FET performance benchmarking. Therefore, a lacuna is being developed as analytically derivable dopant activation percentage as the author of this Book has elucidated in Chaps. 1 and 2 of this Book for the case of silicon material taking assistance from an important reference article cited in Chaps. 1 and 2 of this Book, have not been systematically and computably determined for other materials of focus and pursuit than silicon that we see today for FET device performance improvements. For substrate resistivity as is reported for n and p type silicon at T = 300 K is different from Hall resistivity where the factor mobility μ is now function of both ionized impurity scattering, neutral impurity scattering and for degenerate doping conditions, carrier-carrier scattering and hence the overall scattering time in Drude equation needs to be used for majority carrier n-type mobility and majority carrier p-type mobility calculations. As have been shown, presence of neutral dopants decreases the mobility for majority carrier types in silicon material at T = 300 K, so for better optimized resistivity so that there is minimal voltage drop between the depletion region edge and substrate contact, low non degenerate doping needs to be used in today's GAA nanowire and vertical GAA nanosheet FET architecture.

For cryogenic GAA nanowire FET and vertical GAA nanosheet FET, between the thin body lower edge and oxide isolated substrate contact, due to severe incomplete ionization, the majority activated dopant concentration is substantially reduced for moderate to low non degenerate substrate doping and considerable concentrations of neutral dopants also decrease the majority carrier mobility. Hence the substrate conductivity can be very low and resistivity can be very high at cryogenic temperature than T = 300 K operation. As a result of which, from the thin body of GAA and nanosheet stack to the substrate zero potential, there is high enough resistivity in the bulk to elevate the bottom potential of the thin silicon body to positive potential. Therefore, the mobile holes in the depletion region of thin silicon body n-FET, get repelled by this elevated bottom potential of the thin silicon body and these mobile holes stay in the depletion region and reduce the number of activated acceptor dopants further. The vertical gate field strength that terminates on thin body negatively charged acceptors is of lower value now and it decreases the quantum confinement effect. Decrease of quantum confinement means that sub band splitting results in closely lying states including the minima energy and on one hand inversion layer screening of dopants is increased at larger gate voltage by occupying these closely lying energy states of the sub bands in the conduction band but also inter sub band optical scattering is increased. The positive potential due to increased substrate resistivity of the thin body bottom edge also draws the inversion charge away from the surface, so surface roughness and boundary layer scattering is decreased, but more inversion charge screening and increase of inter sub band scattering where threshold voltage is reduced slightly due to bulk potential decrease as a result of less ionized dopants in the depletion region, may be another factor for self-heating effect (SHE) enhancement of both GAA nanowire FET and vertically stacked GAA nanosheet FET which probably has not been analyzed and taken into consideration in device of modeling of these advanced node FETs at cryogenic temperature operation.

From 3D band structure of conduction band and valence band, deducible formulas have to be generated with parameters in the analytical equations for density of states effective mass of electron and hole, conductivity effective mass of electron and hole from DFT and full band 3D quantum simulation based extractions of these masses for substrate materials other than silicon with which n-FET and p-FET are fabricated. Simulation based peak inversion channel mobility determined by thermally limited phonon scattering as a function of doping density at low lateral drift field can be extracted carrier drift velocity versus lateral field plot with the slope of the curve in the linear region, for the vertical gate field that ensures only the dominant thermally limited phonon scattering, gives the channel mobility. DFT, full band 3D quantum simulation and Ensemble Monte Carlo methods all can be applied to generate this drift velocity versus lateral field data where the lateral field is function of spatially varying channel potential for varying drain voltage. Ensemble Monte Carlo simulation method can be additionally applied to determine the thermally limited phonon scattering time for both n-FET and p-FET after finding the channel mobility peak at a particular substrate doping at T = 300 K. Relating the Drude equation for peak mobility with only dominant scattering event being thermally limited phonon scattering now gives the conductivity effective masses of electron and hole as majority and minority carriers as a function of substrate doping density from where the non parabolicity effect on conductivity effective masses can also be determined for extreme doping concentration values in the vicinity of $10^{20}/cm^3$ to $10^{21}/cm^3$.

Precise determination of activated dopants percentage in the high degenerately doped source and drain junctions near $10^{20}/cm^3$ to $10^{21}/cm^3$ of n and p-FET in today's advanced node architectures is important from contact resistance minimization point of view for FET devices fabricated in silicon and other materials and operated at T = 300 K and for drive current, mobility and saturated drift velocity advantage near cryogenic temperature operation. For T = 300 K, the Chaps. 1 and 2 of this book show that ionization percentage remains almost pinned to 100% for high majority carrier substrate doping near 10^{20}/cm^3 and same is true for 10^{21}/cm^3 away from maximum dopant concentration limit sufficiently away from 5×10^{22}/cm^3 limit from silicon diamond crystal structure. Since 100% ionization in silicon is retained not only at T = 300 K but also down to cryogenic temperature for doping close to 10^{21}/cm^3 for n$^+$ source and drain junctions in n-FET and p$^+$ source and drain junctions in p-FET, near ohmic contact with metal electrode can be ensured for silicon FET for highly low contact resistance. But what about FET built in other substrates with degenerately doped source and drain junctions for n-FET and p-FET? Numerical percentage data of activated dopants in the source and drain junctions for n-FET for high degenerate doping at T = 300 K down to T = 4.2 K for FETs built with other materials than silicon are not available and reported either through DFT and full band 3D quantum simulation or by analytical equations as shown in Chaps. 1 and 2 of this book for silicon. If this activated dopant percentage for degenerately doped source and drain junctions fall near 93–95% for other materials excepting silicon material, source and drain contact with metal in part will be Schottky type and hence will introduce significant contact resistance elevation, decreasing the drive current in saturation bias for logic applications as per Roadmap for the most advanced architecture vertically stacked GAA nanosheet FET in silicon from T = 300 K down to T = 4.2 K. So, it is highly imperative that along with degeneracy induced non parabolicity effect, the actual ionization percentage is computed and reported for group IV materials other than silicon, III–V materials, 2D materials, graphene, carbon nanotube, black phosphorous, hexagonal boron nitride for degenerate doping values of the substrate at T = 300 K down to cryogenic temperature and like in silicon as reported in Table and Plot in Chaps. 1 and 2 of this book, some ranges of non degenerate doping values where the activated dopants in T = 300 K for silicon is found to be going through a minima and sufficiently less than 100% and this should be also checked one by one for every other substrate material by simulation and analytical equations based derivations.

For the cryogenic temperature operation, conductivity effective masses of electron and hole as majority and minority carriers for n-FET and p-FET have to be analytically formulated as a function of temperature as shown in Chap. 5 of this book and it has to be done for all other materials with which some of the high performance n-FET and p-FET devices are fabricated. Scattering times calculations are done by Monte Carlo simulation as a function of temperature or they can be computed by Drude mobility equation at a particular temperature from channel mobility as a function of temperature data, once the conductivity effective mass for that carrier at that temperature is determined. This process has to be performed on all materials other than silicon with which today's n-FET and p-FET devices are fabricated. For ensuring volume inversion and ballistic transport for high drive current in these advanced FET architectures like vertical GAA nanosheet and vertically stacked complementary FET, the 2D scaling length must be minimized at high enough gate and drain voltage with increasing quantum confinement. This will ensure reduced surface roughness scattering, boundary layer scattering and inter sub-band optical and non optical phonon scattering all of which contribute to carrier temperature rise and self-heating effect (SHE) aggravation. With the limits of precision nanofabrication technology available today, instances of random dopant fluctuations which increase with build up of neutral dopant concentration at cryogenic temperature, line edge roughness, line width roughness need to be controlled with accurate 3 sigma standard deviation control on mean threshold voltage variation and critical dimension (CD) variation intra-die and from die-to-die.

References

1. Advanced Semiconductor Fundamentals, Robert F. Pierret. Volume VI, Second edition, Pearson Education Inc., 2003.
2. Semiconductor Device Fundamentals, Robert F. Pierret, Addison-Wesley Publication Company Inc, 1996.